Data Driven Guide to the Analysis of X-ray Photoelectron Spectra using *RxpsG*

This book provides a theoretical background to X-ray photoelectron spectroscopy (XPS) and a practical guide to the analysis of the XPS spectra using the *RxpsG* software, a powerful tool for XPS analysis. Although there are several publications and books illustrating the theory behind XPS and the origin of the spectral feature, this book provides an additional practical introduction to the use of *RxpsG*. It illustrates how to use the *RxpsG* software to perform specific key operations, with figures and examples which readers can reproduce themselves.

The book contains a list of theoretical sections explaining the appearance of the various spectral features (*core-lines*, *Auger* components, valence bands, loss features, etc.). They are accompanied by practical steps, so readers can learn how to analyze specific spectral features using the various functions of the *RxpsG* software. This book is a useful guide for researchers in physics, chemistry, and material science who are looking to begin using XPS, in addition to experienced researchers who want to learn how to use *RxpsG*.

In the digital format, the spectral data and step-by-step indications are provided to reproduce the examples given in the textbook.

RxpsG is a free software for the spectral analysis. Readers can find the installation information and download the package from https://github.com/GSperanza/ website.

RxpsG was developed mainly by Giorgio Speranza with the help of his colleague dr. Roberto Canteri working at Fondazione Bruno Kessler.

Key Features:

- Simplifies the use of *RxpsG*, how it works, and its applications.
- Demonstrates *RxpsG* using a reproduction of the graphical interface of *RxpsG*, showing the steps needed to perform a specific task and the effect on the XPS spectra.
- Accessible to readers without any prior experience using the *RxpsG* software.

Giorgio Speranza is Senior Researcher at Fondazione Bruno Kessler – Trento Italy, Associate Member of the Italian National Council of Research, and Associate Member of the Department of Industrial Engineering at the University of Trento, Italy.

Data Driven Guide to the Analysis of X-ray Photoelectron Spectra using *RxpsG*

Giorgio Speranza

CRC Press
Taylor & Francis Group
Boca Raton London New York

CRC Press is an imprint of the
Taylor & Francis Group, an **informa** business

First edition published 2024
by CRC Press
2385 NW Executive Center Dr, Suite 320, Boca Raton, FL 33431

and by CRC Press
4 Park Square, Milton Park, Abingdon, Oxon, OX14 4RN

CRC Press is an imprint of Taylor & Francis Group, LLC

© 2024 Giorgio Speranza

Library of Congress Cataloging-in-Publication Data
Names: Speranza, Giorgio, author.
Title: Data-driven guide to the analysis of X-ray photoelectron spectra using *RxpsG* /
    Giorgio Speranza.
Description: First edition. I Boca Raton, FL : CRC Press, 2024. I Includes
    bibliographical references. I Summary: "This book provides a theoretical background
    to X-ray Photoelectron Spectroscopy (XPS) and a practical guide to the analysis of
    the XPS spectra using the RxpsG software, a powerful tool for XPS analysis.
    Although there are several publications and books illustrating the theory behind X-ray
    Photoelectron Spectroscopy and the origin of the spectral feature, this book provides
    an additional practical introduction to the use of RxspG. It illustrates how to use
    the RxpsG software to perform specific key operations, with figures and examples
    which the reader can reproduce themselves. The book contains a list of theoretical
    sections explaining the appearance of the various spectral features (core-lines, Auger
    components, valence bands, loss features etc). They are accompanied by practical
    steps so readers can learn how to analyze specific spectral features using the various
    functions of the RxpsG software. This book is useful guide for researchers in physics,
    chemistry, and materials science who are looking to begin using X-ray Photoelectron
    Spectroscopy, in addition to experienced researchers who want to learn how to use
    RxpsG"—Provided by publisher.
Identifiers: LCCN 2023024086 I ISBN 9781032273600 (hardback) I ISBN
    9781032284712 (paperback) I ISBN 9781003296973 (ebook)
Subjects: LCSH: X-ray photoelectron spectroscopy. I RxpsG.
Classification: LCC QC454.P48 S64 2024 I DDC 537.5/35202854678—dc23/
    eng/20231005
LC record available at https://lccn.loc.gov/2023024086

ISBN: 9781032273600 (hbk)
ISBN: 9781032284712 (pbk)
ISBN: 9781003296973 (ebk)

DOI: 10.1201/9781003296973

Typeset in Times
by Apex CoVantage, LLC

To Carla,
walking companion throughout the life

Contents

Preface

This book focuses on the use of *RxpsG*, an open-source software oriented to the analysis of photoelectron spectra. *RxpsG* is based on the R platform, a very popular set of libraries for statistical analysis and graphics. *RxpsG* uses the R platform to generate all the functions needed for spectral data reduction. This book briefly illustrates the XPS technique, describes the various spectral features, and also describes how, for each one, the spectral analysis can be performed using *RxpsG*.

What This Book Is About?

X-ray photoelectron spectroscopy (XPS) is a surface-sensitive technique utilized to analyze the outer few top layers of the materials. This book introduces XPS, as pioneered by Siegbahn and collaborators, providing the basis of this technique and how spectra are generated.

The power to provide the elemental composition and a quantitative chemical analysis, to describe the material's electronic and structural properties, makes XPS an ever-green and broadly diffused tool. However, the usefulness of XPS is directly linked to the ability of obtaining information from the spectra. This book describes the open-source *RxpsG* software, which is oriented to the XPS spectral analysis.

The book is organized in sections dedicated to XPS and the various spectral components provided by this technique.

Chapter 1 gives historical notes about the photoelectron spectroscopy and then turns to the present and future applications of this technique. Afterward, it follows a description of the basis of the photoelectron spectroscopy starting from the Einstein equation of the photoelectric effect. In this chapter, all the theoretical information needed to describe the photoelectron spectra will be given. The photoemission and the equations governing this process as well as the measure of the electron kinetic energy required to obtain the photoelectron spectra will be illustrated. Calibration of the instruments is essential to represent data on an energy scale. Then, effects characterizing the photoelectron spectroscopy as the attenuation of the signal intensity with the depth and the consequent surface sensitivity and the occurrence of spin-orbit splitting will be described. Next, some notes about the sample preparation, the occurrence of contamination, and possible solutions to limit their effects are given. Finally, the main parts of an XPS instrument are illustrated.

Chapter 2 is dedicated to the R platform and the *RxpsG* software. This chapter first describes how the *RxpsG* software can be installed together with base information about the R package. Then it describes how the *RxpsG* software is organized and its main functions.

Chapter 3 explains how to perform a spectral acquisition. It illustrates also the case of non-conducting samples and how spectra can be acquired using the charge compensation. In this case, an alignment of the energy scale to a reference for a correct interpretation of the spectra is needed. This chapter provides information on how to perform this operation in *RxpsG*.

Then, it follows a description of the various types of spectra generated by an XPS instrument: the wide spectra, their main spectral constituents, the identification of various elements, and how this is performed in *RxpsG*. It may be useful to extract spectral features from a wide spectrum and then proceed with a detailed spectral analysis. This function is implemented in *RxpsG* and will also be explained. Then the chapter focuses on the *core-line* spectra, the associated cross section, the splitting in the spin-orbit components, and the secondary spectral structures. This section presents the core-line spectral analysis, as it is carried out in *RxpsG* with a detailed description of the software functions. Then the chapter proceeds with the description of the *Auger* spectra of the valence bands and of the loss features and how they are analyzed in *RxpsG*.

Chapter 4 is dedicated to the quantification and to the elements which can affect it. This chapter will illustrate the factor analysis which is implemented in the R platform. Finally, the chapter gives a description of the functions related to the quantification and reporting of the results.

Chapter 5 illustrates the use of the XPS to obtain information about the depth distribution of elements. This is accomplished via non-destructive depth profiling using the Angle Resolved XPS, or energy resolved XPS, or using the destructive depth profiling performed with surface erosion via ion sputtering.

Chapter 6 illustrates the various graphic options to present and compare the analyzed spectra, customizing the plot according to personal needs. Other useful functions to handle the spectral data are described as well.

Finally, *Chapter 7* collects all the XPS spectra in Ascii format, to reproduce all the examples shown throughout the book, thus helping the reader to better understand the use of the *RxpsG* software.

To Whom This Book Is Addressed?

This book should be of interest to the XPS community needing a tool for the spectral analysis. For scientists who are new to the XPS, this book is a good beginning, which couples theory and practice described in an easy and comprehensive way. *Chapter 1* provides an overview of the basis of the XPS technique, the main spectral characteristics, and short description of a typical XPS instrument. It is useful for beginners who are not experienced with XPS. *Chapter 2* introduces the *RxpsG* software and how it can be installed on a personal computer. It is aimed to all who want to use *RxpsG*. *Chapter 3, 4* and *5* illustrate how XPS spectra can be analyzed using the various *RxpsG* functionalities and how the elemental quantification and depth profiling by ARXPS may be performed. These chapters are useful for beginners and also for experienced scientists who desire to use a new customizable tool for the XPS analysis. This section of the book includes tutorial parts addressed to non-experts and practical parts for all the readers.

Chapter 6 is dedicated to graphics with practical parts and essential information related to the use of the software. *Chapter 7* is a data repository useful for all who want to reproduce the examples shown in this book.

In conclusion, *RxpsG* is an open-source software allowing R users to access the *RxpsG* routines and modify the code to better satisfy their own needs. The description of the data structure enables adding *RxpsG*-compatible functions to perform specific data processing still not included in the package. Finally, this book is addressed to all those using XPS and moved by the desire to create a community of *RxpsG* users which could further develop and improve the software.

Version Information

The book describes the *RxpsG.2.3-2* version of the software. The package was built using the R 3.4.2 version. *RxpsG* should be compatible with newer versions of R. It can be easily recompiled as indicated in Chapter 2 with newer versions of R. The recompilation of the *RxpsG* package is needed also in the cases when the macros are modified or added to customize the software, to render the changes permanent. The R base platform can be downloaded from R-CRAN website www.R-project.org/, where one can find latest versions. This website also provides the most up-to-date information as well as helps pages which can be useful not only to better work with *RxpsG* but also to take profit of all the potentialities of the R libraries for multipurpose data analysis.

Acknowledgments

The *RxpsG* package was inspired by my friend and colleague Dr. Roberto Canteri, who, at that time, was already expert in R programming. He finally convinced me to develop this tool for the analysis of the photoelectron spectra and, in principle, any kind of data. He developed the *RxpsG* core and the base procedures allowing all the various functions to work correctly. Thank you Roberto also for the constructive discussions and the help for writing the correct codes.

I would like to thank also Prof. John Verzani of The City University of New York for the gWidgets2 and gWidgets2tcltk packages I deeply utilized to develop all the *RxpsG* graphical user interfaces (GUIs) and for his help and substantial solutions to complex problems.

Finally, I would like to thank the R Core Team for the open-source R platform for all the effort to make it efficient and stable and thanks to the R community for developing the R libraries and maintaining them always up-to-date.

Giorgio Speranza
Trento, Italy.

1 Introduction to Surface Analysis by X-ray Photoelectron Spectroscopy

1.1. INTRODUCTION

The physical and chemical properties of a material are tightly related to electronic structure of the component atoms. Crystalline or amorphous phases and the material chemical composition strongly influence the charge distribution around the nuclei and the electric potential they generate. Then, a precise description of the material properties requires the ability to accurately probe the electronic structure of atoms and how it changes upon the different chemical configurations. X-ray photoelectron spectroscopy (XPS) utilizes X-photons to investigate the material properties. The term X-ray is used to indicate a radiation with wavelengths in the range 10–0.01 nm corresponding to soft and hard rays. Different frequencies of X-photons correspond to the different kinds of light/matter interactions. This led to the development of a number of techniques to probe materials at different length scales, from the macroscopic dimensions to the atomic level: in the first case, bulk structural and chemical information are provided (X-ray diffraction, X-ray fluorescence, X-ray tomography, etc.), while in the second case, the description of the local environment of the atoms is obtained (Extended X-ray Absorption Fine Structure and X-ray Absorption Near-Edge Spectroscopy, X-ray Photoelectron Spectroscopy).

This book focuses on the use of XPS to characterize the material's properties. The ability to exactly describe the characteristics of a given solid depends on the proficiency in correctly utilizing the XPS instrument and on the skill to extract the information from the XPS spectra. This book is aimed at describing a tool for analyzing the XPS spectra. It is directed to a wide community composed by both beginners and experts. The book contains basic information and tutorials for who is new with the XPS technique. These parts are coupled to practical sections illustrating the use of the *RxpsG* software which can be interesting for experienced scientists looking for a customizable tool for the spectral analysis.

Chapter 1 introduces the reader to the XPS, providing a short history of this technique. Then, the base principles of this analytical tool are illustrated and then it follows a description of the scattering processes associated with the photoemission in the material bulk and a recap of the j-j and L-S coupling needed to describe the main characteristics of the core-line spectra produced by XPS. Chapter 1 includes also some notes regarding the sample preparation, since this may strongly affect the quality of the results, and the chapter concludes with a synthetic description of a typical XPS instrumentation.

DOI: 10.1201/9781003296973-1

1.2. X-RAY PHOTOELECTRON SPECTROSCOPY: HISTORICAL NOTES

XPS took its first steps at the end of 18th century with the experiments of Helmoltz, who observed that pieces of metal under vacuum were sparking when irradiated with UV light [1]. These experiments gave rise to an intense research leading to the derivation of the ratio h/e, where h is Plank's constant and e is the electron charge. In particular, Lenard in his experiments observed that the kinetic energy of photoelectrons from a given emitter was dependent not on the light intensity but on the light frequency [2]. In addition, he was the first to correctly account for the work function that could not be understood by the classical wave theory of light. The generation of electric charge upon irradiation was exhaustively explained by Einstein, in 1905, who described the photoelectric effect as the result of the energy transfer from photons to electrons. His papers filled the gap between the energy properties of light and statistical description of gas with the equation

$$E = R/N \cdot \beta v \qquad\qquad 1.1$$

Here, $R/N = k$ is the Boltzmann constant and β is the radiation constant in the Planck formula for the energy density of black-body radiation [3]. Equation (1.1) is equivalent to $E = hv$ introduced by Millikan [4], where $h = R\beta/N$. It is worth noting that this concept was revolutionary. At the beginning of the 1900s, the relation between Planck's quantum theory and the light photons was still not accepted by the whole scientific community and still some physicists rejected the concept of light quanta.

Then, considering the energy conservation principle, Einstein explained the photoelectric effect as a result of irradiation by light quanta. Einstein assumed that every electron spends a characteristic work Φ to leave a solid. Then reconsidering equation (1.1), the kinetic energy of such electron can be described as [5–7]

$$1/2\, mv^2 = E_k = hv_0 - \Phi \qquad\qquad 1.2$$

where Φ represents the electron work function, the minimum energy required to remove an electron from a solid. Changing the photon energy, it was possible to estimate the work function (Φ) of the respective material measuring the kinetic energy E_k. Meanwhile, Robinson and Rawlinson in 1914 were the first scientists to observe photoemission from X-irradiation [8]. Equation (1.2) was subsequently refined thanks to the work of Rutherford in 1914, closer to the basic XPS equation [9]:

$$E_k = hv_0 - E_B - \Phi \qquad\qquad 1.3$$

which introduces the binding energy of the electron in the solid. The possibility to measure the electron kinetic energy as a function of the photon energy was utilized by Steinhardt and Serfass in 1951 to develop an analytical tool [10]. The merit of these two scientists was the first attempt to translate the simple experiments on the photoelectric effect into an instrument performing the surface chemical analysis. These scientists were able to acquire characteristic spectra from metals such as Cu, Zn, Rh, Ag, and Au, showing that they could make a qualitative analysis of a Cu–Zn alloy and a quantitative analysis of an Ag–Au alloy [10]. However, their spectrometer was not as performing as they are today. Their spectra consisted of a series of bands with reasonably well-defined edges where authors were able to derive spectral features characteristic of the material.

The great advancement in the development of X-ray photoelectron spectroscopy into a reliable analytical technique was made by Kai Siegbahn at the University of Uppsala in Sweden [11–13]. For the research done to develop high-resolution electron spectroscopy for chemical analysis, Siegbahn won the Nobel Prize in 1981. Thanks to his work, this technique ranks as one of the most important in surface science and chemical physics in general. The information supplied by this technique is crucial for understanding the electronic structure of atoms, molecules, and ions in solid materials.

1.3. PRINCIPLES OF THE TECHNIQUE

As seen in the previous section, XPS deals with the photoelectric effect namely the ejection of photoelectrons from atomic orbital induced by an X-photon radiation of energy hv. The emitted photoelectrons are then analyzed by a spectrometer which measures their kinetic energy and the photocurrent intensity. The resulting data are presented as spectra where the intensity (commonly expressed as counts/s) is drawn with respect to the electron kinetic (or binding) energy. Equation (1.3) tells us that the electron kinetic energy E_k measured by the analyzer is dependent on the exciting photon energy which is defined by the X-ray source adopted in the instrumentation. Generally, Mg or more frequently Al anodes are used in modern instruments to produce Kα radiations at 1253.6 eV and 1486.6 eV, respectively. These values define the maximum spectral energy that can be analyzed corresponding to photoemission processes without any energy loss. XPS spectra are recorded on an energy range which typically spans from 0 till 1,200–1,300 eV. Although kinetic energy is measured, frequently the XPS spectra are presented on a binding energy (E_B) scale, because it more directly identifies the electron source, specifically its parent element and the initial atomic energy level. The binding energy is defined by equation (1.3) $E_B = hv - E_K - \Phi_{samp}$. Solid samples introduced in an XPS instrument are in contact with the spectrometer in an attempt to avoid charging effects and maintain a well-defined potential during the analysis. In this simple case, we can represent the sample energy levels and the photoelectron kinetic energies in the diagram of Figure 1.1.

In this scheme, there are two regions: the one referred to the sample and the other referred to the spectrometer. Because the sample and the spectrometer are connected, they are in a thermodynamic equilibrium resulting in the same electron chemical potential, that is, the Fermi level referred to the sample and to the spectrometer are the same. For a conducting sample, as in this example, the Fermi level corresponds to the highest occupied molecular orbital (HOMO). Above the valence band is the conduction band while below are the core levels. Following the scheme, the kinetic energy of a photoelectron is then given by equation (1.3), where the work function Φ_S is defined as the work needed to make the electron to overcome the potential barrier at the sample surface and to reach the vacuum level. However, in a real case, the photoemitted electron is accelerated or retarded (as it is in Figure 1.1) by a potential.

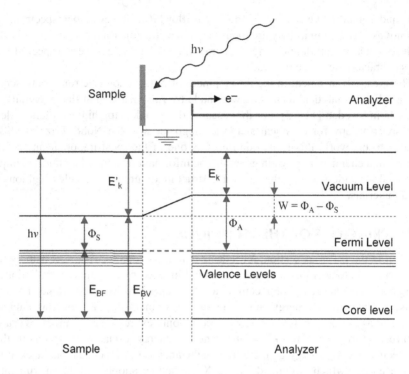

FIGURE 1.1   Diagram of the energy levels for a sample in contact with the electron spectrometer and the instrument chassis which is normally grounded as indicated in this scheme. A generic core level, the valence band, and the Fermi level are indicated. In this scheme, the Fermi level corresponds to the HOMO level as it occurs in metals. E_{BF} is the binding energy associated to a core level and referred to the Fermi level. Similarly, E_{BV} is the binding energy of a core level referred to the vacuum level. A diagram holds also for semiconducting or insulating materials with the only difference that the Fermi level lies somewhere between the valence and the conduction band above.

A spectrometer work function Φ_A is defined as the sum of the sample work function plus an accelerating/retarding potential W

$$\Phi_A = \Phi_S + W \qquad\qquad 1.4$$

As a consequence, the electron kinetic energy really measured by the spectrometer is

$$E_{kin} = h\nu - E_{BF} - \Phi_S - W$$
$$E_{kin} = h\nu - E_{BF} - \Phi_A \qquad\qquad 1.5$$

or

$$h\nu = E_{kin} + E_{BF} + \Phi_A$$

As seen earlier, the binding energy E_{BF} is referred to the Fermi level. What generally is done in XPS instruments is to adjust the Fermi level to fall at 0 eV. Generally, Ag or Au reference samples are selected to define the Fermi Edge. Essentially an XPS instrument allows the reconstruction of the atom's electronic structure as shown in Figure 1.2. A photoelectric peak corresponds to each of the atomic energy levels. The same occurs for the density of states (DOS) near the Fermi level which is reproduced in the valence band. Note that, in the scheme of Figure 1.2, the *Auger* transitions are not indicated.

The working principles of photoelectron spectroscopy are the same as those of the original instruments. Apart from the increased energy resolution, great advancements were made to make the energy analyzers more sensitive and sophisticated. Nowadays, the spectrometers measure not only the photoelectron kinetic energy, but they are also capable to measure their momentum p (wave vector p/ℏ)

$$p = \sqrt{(2m\,E_k)} \qquad\qquad 1.6$$

which corresponds to the detection of electron energy distribution in a certain direction defined by the angles (ψ, θ) with respect to the impinging light. This will be better described in the section Angle-resolved XPS.

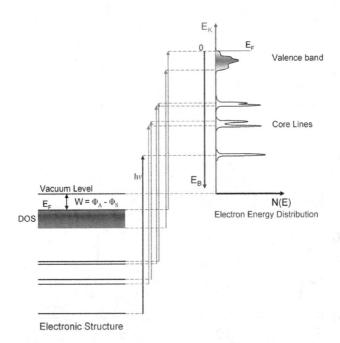

FIGURE 1.2 Relation between the energy levels in a solid and the electron energy distribution obtained via photoemission induced by photons of energy hv. As shown, the photoelectron intensity measured in counts is plotted against their kinetic energy with its zero at the vacuum level of the sample $E_{kin} = hv - E_{BF} - \Phi_A$. The maximum kinetic energy corresponds to E_F at binding energy = 0.

1.3.1. INSTRUMENT CALIBRATION

Calibration of the instrument energy scale means finding a stable energy reference valid for any kind of material. For clarity, let us consider the binding energy as defined by equation (1.5) instead of the kinetic energy. In conducting samples, as metallic samples, the binding energy E_B of the photoemitted electrons is measured with respect to the Fermi level which is assumed to be at $E_B = 0$ eV. This corresponds to the maximum kinetic energy $h\nu - \Phi_A$ as shown in Figure 1.2. Generally the position of the Fermi level is defined in metals as Ag or Au. Metals have the peculiar properties that the density of state near the Fermi level possesses a step-like trend. An example of the silver valence band (VB) is shown in Figure 1.3a while Figure 1.3b shows the expanded portion of the VB near the Fermi Edge. The position of the Fermi level is assigned to the midpoint between the minimum and maximum levels of the spectrum in this region.

Because the noise is superposed to the spectrum, different methods are applied to find the Fermi level. For example, the midpoint is defined selecting the spectral region between a certain percentage (84%–16%) of the average spectral intensity

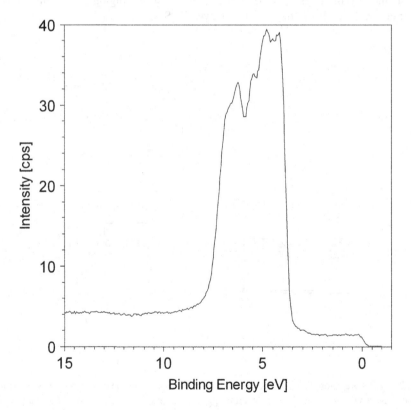

FIGURE 1.3   (a) Valence band of silver and (b) expanded view of the region near the Fermi Edge. It is shown also the valence band fitting with the Fermi-Dirac function while the star represents the position of the Fermi level derived from the fit.

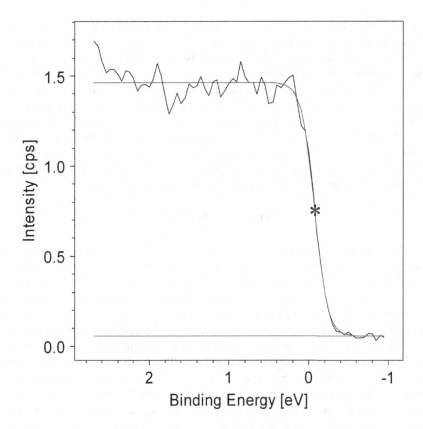

FIGURE 1.3   (Continued)

before and after the signal decay. Another possibility is to fit the VB using the function describing the Fermi-Dirac distribution [14]. This function is defined by

$$N(E) = 1/\left[\exp\left(E - E_F\right) / K_B T + 1\right] \qquad 1.7$$

This function can be utilized to fit the upper part of the valence band and to derive the position of the Fermi level. The calibration of the instrument is then made forcing this point at zero eV on the binding energy scale as in Figure 1.3.

Another possibility is to acquire the core-lines of reference materials falling at a well-tabulated position. Because the value of their binding energy is referred to $E_F = 0$, it is possible to calibrate the binding energy scale. Examples are the Ag $3d_{5/2}$ which must be at 368.2 eV, or the Au $4f_{7/2}$ must fall at 84.0 eV. Deviations from these values derive from non-calibrated position of E_F [15].

The instrument calibration includes also a second step consisting in checking the linearity of the analyzer response. The measure of the photoelectron energies from different elements is based on the assumption that the energy analyzer has a linear response over an extended binding energy scale. If this is the case, the spectra of reference materials located in different energy regions of the BE scale should be obtained in the correct position. At this scope, Au 4f, Ag 3d, and Cu 2p core-lines are utilized, since they fall, respectively, at 84 eV, 368.2 eV, and 932.6 eV, that is, at beginning, middle, and near to the end of the BE scale. The acquisition of the indicated core-lines from sputter-cleaned samples thus allows calibrating the analyzer. In modern instruments, automated procedures are generally available to carry out the analyzer calibration. A final note concerns the calibration the spectrometer transfer function. This function describes the detector performances as a function of the photoelectron energy. The detector (array of channeltrons, 2D multichannel plate, or other) response is different for different electron kinetic energies. Generally it is higher at higher kinetic energies. It is then needed to correct the spectral intensity I(E) to make the detector response constant. I(E) can be described as

$$I\big(E\big) = I_0 Q\big(E\big) n\big(E\big) \qquad\qquad 1.8$$

where I_0 represents the primary X-ray beam flux, Q(E) is a function representing the characteristic of the instrument analyzer for given settings, and n(E) is the spectrum emitted from sample. Assuming fixed the instrumental settings such as pass energy, slit widths, and sample position, then Q(E) will be the same for all the spectra, as it is independent of the sample used. However, in normal practice, the instrumental parameters cannot be kept constant and must be adapted to optimize the acquisition. So Q(E) may vary in a way dependent on the instrument manufacturer, and Q(E) must be characterized for all the acquisition conditions. This has been done at the National Physics Laboratory (NPL), thus offering the possibility to estimate n(E) [16]. From equation (1.8) follows that possessing a reference sample with known n(E), then I(E) may be measured and Q(E) can be obtained for any acquisition condition. This permits to estimate the true n(E) spectrum for a generic sample.

1.3.2. PHOTOEMISSION: ATTENUATION LENGTH AND INELASTIC MEAN-FREE PATH

X or UV photons may penetrate deeply in the bulk of the specimen generating photoelectrons with maximum energy corresponding to that of the exciting photon. However, experiments show that the probability that photoelectrons generated below the surface will leave the solid with their original energy is rather low. The photoelectrons produced in the sample bulk can travel only a finite distance before they scatter either elastically or inelastically. For these two processes is defined an average distance an electron travels through a condensed phase between scattering events. This corresponds to the elastic mean-free path (EMFP) λ_e in the case of elastic scattering, and to the inelastic mean-free path (IMFP) λ in the other case. For random scattering, calculations suggest, that in crystalline materials, $\lambda_e < \lambda$ [17]. However, there is a degree of uncertainty in modeling elastic, quasi elastic (phonon mediated), and inelastic scattering processes in the 0–1,500 eV range typical of XPS. Consequently the contributions of elastic scattering to electron attenuation are often neglected in models describing the mean-free path of electrons in a material.

When inelastic scattering occurs, the photoelectron loses part of its original kinetic energy. The maximum distance traveled by the photoelectron is determined by its initial kinetic energy before it is completely dissipated in successive inelastic scattering processes. Consequently, the sampling depth depends on the photon energy and on the amount of energy transferred to the electron which is described by equation (1.3). This allows probing the materials at different depths by varying the photon frequency as done in synchrotron radiation sources. The scattering processes then limit the number of photoelectrons reaching the surface with enough energy to leave the material, thus causing an attenuation of the electron signal with the depth. Let us consider photoelectrons created at a depth d below the surface with an angular distribution depending on the orbital involved. Because energy losses are due to inelastic scattering, the signal attenuation is described by the same experimental expression which holds for the light attenuation in an absorbing medium, the Beer-Lambert law:

$$I = I_0 \exp\left(-d / \lambda \sin\theta\right) \qquad\qquad 1.9$$

where I is the attenuated signal intensity, I_0 is the non-attenuated electron current as it were generated on the surface, and θ is the electron "take-off" angle defined by the electron trajectory and the sample surface. Note that generally the tilt angle α defined as the deviation of the electron trajectory from the vertical analyzer axis ($\alpha = 90 - \theta$) is used. Finally, λ represents IMFP of the photoelectron at a given energy. The inelastic mean-free path may be evaluated on the basis of the nature of the material analyzed [18, 19]. In particular, it may be determined by measuring the attenuation of a characteristic electron signal by an overlayer of known thickness.

For a sample consisting of an overlayer A with thickness d, and a bulk substrate B, assumed A and B both of uniform composition, equation (1.9) leads to the two relations

$$I_A = I_A^\infty \left[1 - \exp\left(-d / \lambda_A \sin\theta\right)\right] \qquad\qquad 1.10$$

$$I_B = I_B^\infty \exp\left(-d / \lambda_B \sin\theta\right) \qquad\qquad 1.11$$

I_A and I_B are the measured intensities, I_A^∞ and I_B^∞ are the intensities from A and B bulk samples measured under identical conditions. Finally, λ_A and λ_B are the inelastic mean-free paths of electrons relative to measured core-lines of A and B traveling through the overlayer A. Consider that λ_A and λ_B are different because of the different electron kinetic energies of electrons emitted by A and B atoms and because of the different material densities ρ_A and ρ_B. Equations (1.10) and (1.11) were utilized to experimentally estimate the inelastic mean-free path of the material A with known thickness d.

If we consider equation (1.11), the mean distance <z> between two consecutive collisions may be expressed by equation (1.12):

$$<z> = 1/I_0 \int_0^\infty z \left(N(z)/\lambda\right) dz = \lambda \qquad\qquad 1.12$$

Then λ represents the mean distance traveled by an electron with a given energy between successive inelastic collisions. From equation (1.11), it is also possible to evaluate the probability $P(z)$ that an electron with energy E_0 will move for a path z without experiencing inelastic scattering:

$$P(z) = 1/I_0 \int_0^\infty I_0 \exp(-z/\lambda) = \exp(-z/\lambda) \qquad 1.13$$

From equation (1.13) follows that the probability that an electron covers a distance equals to λ in the sample without undergoing inelastic collisions is $e^{-1} = 0.368$. This led to the definition of the *escape depth*. Assumed an exponential decay of the electron signal, the escape depth d is defined as the depth at which the signal drops by a factor e^{-1} (36.8%) of its original value. This corresponds to the 95% of probability that an electron escaping without significant energy loss (due to inelastic scattering process) will come from a depth of three times d. Considering possible presence of sample inclination as illustrated in Figure 1.4, then for a take-off angle θ and an attenuation length λ, the escape depth $d = \lambda \sin\theta$. Because the λ values are derived from the measurement of the "attenuation" of the signal intensity, they are frequently referred to as *attenuation lengths* λ_{AL} although there is not a well-established definition of attenuation length because of the effect of elastic scattering.

It is possible to define an average value for the *mean escape depth* as

$$< d > = \int_0^\infty z\, \Phi(z, \theta)\, dz / \int_0^\infty \Phi(z, \theta)\, dz \qquad 1.14$$

where $\Phi(z, \theta)$ represents the emission distribution function of the material at depth z from the surface and at an emission angle θ with respect to the normal to the sample surface. If the effects of the elastic scattering are ignored, <d> corresponds to IMFP multiplied by $\cos\theta$.

<d> offers the possibility to estimate the surface sensitivity, because it strongly depends on the electron energy and on the nature of the sample analyzed. From the

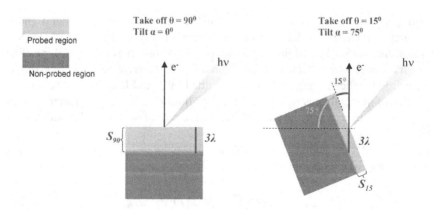

FIGURE 1.4 Sampling depth S and attenuation length λ are indicated in two cases: take-off angle = 90° and 15°. As it can be seen at 15°, the real sampling depth S_{15} with respect to the sample surface is much lower than $S_{90} = 3\lambda$.

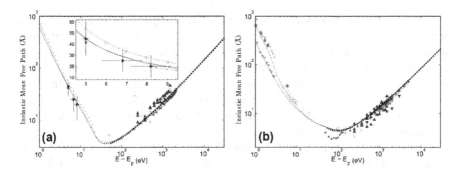

FIGURE 1.5   Electron IMFP (a) for Al and (b) for Au. Experimental data are modeled with a Penn algorithm; the solid line and the dashed line correspond to different fitting parameters. Different symbols represent different experimental data as indicated in [23].

Source: Reprinted with permission from [23]

previous notes, the information obtained from the depth x below the surface will be a percentage of the total information given by

$$< d > \ = \int_0^x \Phi\left(z, \, \theta\right) dz \Big/ \int_0^\infty \Phi\left(z, \, \theta\right) dz$$

$$= \int_0^x \exp(-z\,/\,\lambda)\,dz \Big/ \int_0^\infty \exp(-z\,/\,\lambda)\,dz = 1 - \exp(-z\,/\,\lambda) \qquad 1.15$$

From equation (1.15) follows that at a depth $3d = 3\lambda_{AL}\sin\theta$, the 95% of emitted electrons are detected. This value is defined as the *sampling depth* also called information depth. The sampling depth corresponds to the maximum depth normal to the surface from which sample information can be obtained. Increasing the sample inclination, the sampling depth will reduce and the analysis will become more superficial.

The sources of attenuation length were studied in the past by Seah and Dench [20] collecting experimental attenuation length values obtained from overlayers deposited on substrates by applying equations (1.10) and (1.11). λ_{AL} values expressed in monolayers were obtained modeling the data using a "universal curve". Authors fitted the values of the attenuation length separately for elements and organic and inorganic compounds [21] and [22].

Results for Al and Au are shown in Figure 1.5.

1.3.3.   J-J AND L-S COUPLINGS AND SPIN-ORBIT SPLITTING

The electron is described by three quantum numbers: $n = 1, 2, 3 \ldots$ is the principal quantum number indicating the energy level, $l = 0, 1, 2, 3 \ldots$ and $s = \pm \frac{1}{2}$ represent the orbital angular momentum and the spin associated to the electron. Possessing a charge, it generates a magnetic field moving around the nucleus. The intensity of the whole magnetic field depends on the value of the total angular momentum J due to all electrons orbiting around the nucleus. To determine the value of the total angular momentum, one possibility is to compute the single angular momentum $j_i = l_i + s_i$ for each electron i. Then J is given by $\Sigma_i j_i$. This description of the total orbital momentum obtained by summing individual j_i values is called *j-j* coupling. This description is suitable for atoms with atomic number ≥ 75. The *j-j* coupling is at the basis of the X-ray notation used to describe the atomic orbitals involved in the X-ray generation or the Auger transitions. In this notation, $n = 1, 2, 3, 4 \ldots$ are indicated with the letters K, L, M, N $\ldots$ while the values of l and j are indicated by the suffixes according to those reported in Table 1.1.

Another way to obtain the total momentum J is to compute the total orbital angular momentum L as summation of all the individual values of l_i, $L = \Sigma_i l_i$ and the total spin momentum S as a summation of the individual spin s_i, $S = \Sigma_i s_i$. Then the total momentum J is then obtained by coupling L and S. Because S can be parallel or anti-parallel to L, the vectorial summation will be $J = |L \pm S|$. This second description of the total momentum is called *L-S* coupling (or Russel Saunders coupling) and was found to be effective to describe atoms with atomic number $Z \leq 30$. In this scheme, the annotation is in the form $^{(2S+1)}L$ which describes the electron distribution in the final state where the total angular momentum $L = 0, 1, 2, 3$ is indicated by the letters S, P, D, F, etc. and the superscript denotes the multiplicity $2S + 1$. In this annotation, $2J + 1$ is used as subscript to indicate the degeneracy due to the total angular momentum J. In Table 1.2 are listed the terms corresponding to the various L, S, and J values.

Apart from the annotations, the coupling between orbital momentum and spin results in a splitting of the core-line spectrum. This last corresponds to a current generated by the sum of single photoemitted electrons. Then, the spectral properties are linked to those of the parent single electrons. Essentially, for the single electron, the spin $s = \pm \frac{1}{2}$ couples with the orbital angular momentum $l = 0, 1, 2, 3 \ldots$ for *s, p, d,* and *f* orbitals. This coupling leads to different values of j for each of the two

---

TABLE 1.1
X-ray Spectroscopic Notation

Quantum numbers					
n	l	j	X-ray suffix	X-ray level	Spectroscopic level
1	0	1/2	1	K	$1s_{1/2}$
2	0	1/2	1	L_1	$2s_{1/2}$
2	1	1/2	2	L_2	$2p_{1/2}$
2	1	3/2	3	L_3	$2p_{3/2}$
3	0	1/2	1	M_1	$3s_{1/2}$
3	1	1/2	2	M_2	$3p_{1/2}$
3	1	3/2	3	M_3	$3p_{3/2}$
3	2	3/2	4	M_4	$3d_{3/2}$
3	2	5/2	5	M_5	$3d_{5/2}$
...	...	...	...	...	...

TABLE 1.2

Notation in L-S Coupling in Some Electronic Configurations

Configuration	L	S	J	L-S notation	IC notation
$(ns)(n's)$	0	0	0	^1S	$^1S_0$
	0	1	1	^3S	$^3S_1$
$(ns)(n'p)$	1	0	1	^1P	$^1P_1$
	1	1	2	^1P	$^1P_2$
$(np)(n'p)$	0	0	0	^1S	$^1S_0$
	0	1	1	^3S	$^3S_3$
	1	0	1	^1P	$^1P_3$
	1	1	2	^3P	$^3P_5$
	2	0	2	^1D	$^1D_5$
	2	1	3	^3D	$^3D_7$
...	...	...	...	...	...

Note: n and n' represent generic values of the principal quantum number.

TABLE 1.3

Spin-Orbit Parameters and Area Ratio R

Subshell	j values	R
s	1/2	//
p	1/2 3/2	1: 2
d	3/2 5/2	2: 3
f	5/2 7/2	3: 4

spin states which are then characterized by different energy values. This splitting is indicated as *spin-orbit* splitting. The magnitude of the separation between the energy levels is linked to the expectation value $1/r^3$. It is then expected that the separation increases with increasing the atomic number Z for a given subshell (n, l constant) or is expected to increase when l decreases (n constant). Finally, in a spin-orbit doublet, the relative peak intensity is proportional to the respective value of $(2j+1)$ degeneracy as it is summarized in Table 1.3

1.4. SAMPLE PREPARATION NOTES

As seen, the XPS sampling depth is limited to some nanometers. The atoms on the sample surface are those largely contributing to the XPS spectra. Great attention is then needed to ensure a clean sample surface avoiding the presence of exogenous chemical species. The preparation of the sample prior to the analysis is then an important and sometimes complex task needed to ensure meaningful analyses. The preparation of the sample consists in mounting the sample on a sample holder and allowing the sample to be inserted in the instrument and positioned under the analyzer. This operation is commonly carried out wearing glows to minimize contaminations deriving from the skin which could compromise the analysis. Similarly, also all tools used to manipulate the sample should be carefully cleaned to reduce all the source of contamination. Sample cleaning is widely used to limit the degree of contamination on the sample surface. Solvents are used to remove organic contaminants. However, solvent itself may leave residuals. In the case of organic samples, there is no way to recognize the contribution deriving from solvents. One solution to this problem is to use appropriate solvents containing an element which is not part of the sample. This element can be used as a marker to identify/quantify the presence of solvent traces. Depending on the kind of material, particular treatments as heating, Ar^+ sputtering, or sample fracturing are performed under vacuum to remove contaminants or oxides and obtain a highly pure surface. Then the sample is fixed to the sample holder using springs or clips, paying attention that they do not interfere with the analyzed area which can greatly increase if sample tilting is required. Attention has to be paid also if the samples are not conductive. In this case, compensation may be more easily achieved by covering the sample with a thin metal sheet with a small window of the dimension of the analyzed area. Presence of the metal facilitates the charge compensation.

Powders can be a problem because of their volatility, the atmosphere trapped among grains, and the scarce compactness. The high surface area renders the introduction of powders under vacuum rather difficult. The sample degassing may disaggregate the powder grains rendering them volatile. Generally, powders are pressed in recesses of the sample holders in an attempt to render the material more compact. Indium foils may also be used to block the powder grains when pressed on its surface. The use of indium has also the advantage to facilitate the compensation of the surface charge which, in the case of non-conductive powders, may severely affect the spectrum quality.

1.5. XPS INSTRUMENTS

XPS instruments are composed by an X-ray source, an analysis chamber, a sector containing the extraction lenses guiding the photoelectron to an energy analyzer, and a detector. All these blocks are kept under high vacuum in the range of 10^{-9} mbar (apart from instruments working *in operando* conditions where the sample is at nearly ambient pressure). Then pumping stages are needed to maintain this high degree of vacuum. Generally, the analysis chamber, the lenses, and the analyzer are pumped by ionic and titanium sublimation pumps to avoid contaminations. Figure 1.6 shows a schematic for a typical high-resolution XPS instrument. In this scheme, the X-ray source, the quartz crystals used to monochromatize the X-photons, and the analyzed region are placed on a Rowland circle.

This arrangement ensures that both the X-source and the sample surface are in focus of the monochromator. In this geometry, the X-photons produced at the X-ray source Al anode are dispersed on the monochromator crystals. The Al Kα radiation possesses a wavelength which allows monochromatization: the quartz crystalline plains act as mirror reflecting the incident X-rays to the sample and the angle

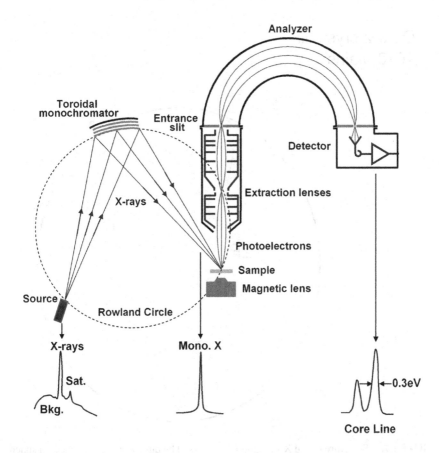

FIGURE 1.6 Scheme of an XPS instrument. The various elements composing the instrument are sketched.

between incident and diffracted rays is 23°. However, to make the Bragg condition to be met at all points, the 10$\bar{1}$0 plains of the monochromator quartz crystals must be bent onto a spherical surface from radius of 2R (outer side of the crystals), while the inner side of the crystals are fine corrected on a curvature of R being R, the radius of the Rowland circle (see Figure 1.7) [24]. According to the dimension of the Rowland circle, the dispersion of the crystals will vary. For a 500 mm Rowland circle, the dispersion corresponds to 1.6 mm/eV.

This geometry allows higher intensities (typically eightfold), higher resolution (typically fourfold), and lower background. The presence of a monochromator results in an X-radiation with narrower monoenergetic X-photons without background and satellites. The X-radiation is focused on the sample surface allowing high photon densities, where they produce a certain electron emission and possible charging needing compensation. However, for non-conducting samples, any surface charging is corrected by using special guns "flooding" the sample surface with low energy electrons compensating the surface charging.

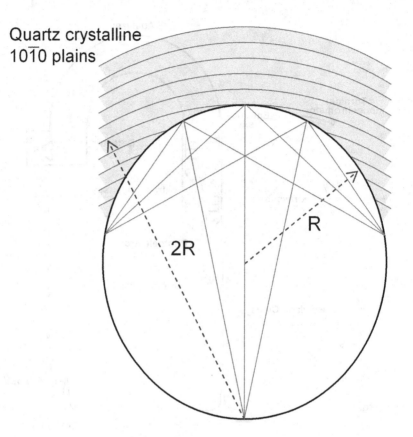

Quartz crystalline
10$\bar{1}$0 plains

2R

R

FIGURE 1.7. Structure of the X-ray monochromator. The quartz crystal is bent on a sphere of radius 2R (R = Rowland radius). Then the inner part of the crystals is finished in such a way to reach a curvature of R. In this way are met the condition of Bragg reflection in any point of the Rowland sphere.

In monochromatized sources, only a fraction of the X-photons, those possessing the correct wavelength, will be diffracted at the correct angle and reach the surface sample. It follows that, for a given value of dissipated power at the Al anode of the X-source, a considerable lower photon density will be available at the specimen with respect unmonochromatized sources. Solution to this problem is increasing the current and acceleration voltage of the X-source or increases the efficiency of the photoelectron collection. In the first case, more powerful sources were produced using rotating anodes to allow power dissipation. On the other side, magnetic lenses were designed to direct photoelectrons toward the entrance iris of the analyzer, where the electrostatic lenses are placed. The typical solid angle of pure electrostatic lenses is ~25° is increased to 90° with the use of magnetic lens resulting in a higher instrument sensitivity. Modern instruments are generally equipped with a magnetic lens guiding the photoelectrons into the lens stage, where they are directed to the analyzer. In spectral mode, only the electrons with trajectories allowed by the extraction lenses are collected. An image of the photoelectrons ejected by the sample surface is projected on a plane at the analyzer entrance slit. Commonly a carousel with a variety of apertures of different sizes allows varying the area of the sample from which electrons are collected (the aperture section divided by the extraction lens magnification defines the extension of the analysis area). The electron energy analyzer consists in a hemispherical condenser deflecting the photoelectrons toward the detector. The deflection depends on the electric force generated by a potential V applied to the condenser. The higher the photoelectron energy, the higher the potential required to direct them to the detector. Entrance and exit slits are utilized to exactly define the photoelectron trajectories for a given potential V allowing an exact measure of the photoelectron energy E_k. Smaller entrance slits are used to increase the energy resolution at expenses of the signal intensity. Electrons arriving to the detector are amplified, and a spectrum reporting the signal intensity as a function of the energy is then obtained. Different kinds of detectors are utilized. Thanks to the special geometry of the hemispherical analyzer, the image projected at the entrance slit may be replicated at the analyzer exit. In this case, a 2D detector may reproduce an image of the photoelectrons emitted by the sample surface for selected energies. This enables chemical maps of the sample surface to be acquired.

REFERENCES

[1]   H.R. Hertz, Ueber einen Einfluss des ultravioletten Lichtes auf die electrische Entladung, Ann. Phys., 267 (1887) 983–1000.

[2]   P. Lenard, Ueber die lichtelektrische Wirkung, Ann. Phys., 313 (1902) 149–198.

[3]   S. Halas, 100 years of work function, Mater. Sci. Pol., 24 (2006) 951–968.

[4]   R.A. Millikan, A direct photoelectric determination of Planck's "h", Phys. Rev., 7 (1916) 355–388.

[5]   W.F. Egelhoff, Core-level binding-energy shifts at surfaces and in solids, Surf. Sci. Rep., 6 (1987) 253–451.

[6]   W. Schattke, M.A. Van Hove, Solid-State Photoemission and Related Methods: Theory and Experiment, Wiley, Weinheim, 2003.

[7]   C. Miron, P. Morin, Handbook of High-resolution Spectroscopy, John Wiley & Sons, New York, 2011.

[8]   H. Robinson, W.F. Rawlinson, The magnetic spectrum of the β rays excited in metals by soft X rays, Philos. Mag., 28 (1914) 277–281.

[9]   E. Rutherford, The connexion between the β and γ ray spectra, Philos. Mag., 6 (1914) 305–319.

[10] R.G. Steinhardt, E.J. Serfass, X-ray photoelectron spectrometer for chemical analysis, Anal. Chem., 23 (1951) 1585–1590.

[11] C. Nordling, E. Sokolowski, K. Siegbahn, Precision method for obtaining absolute values of atomic binding energies, Phys. Rev., 105 (1957) 1676.

[12] E. Sokolowski, C. Nordling, K. Siegbahn, Chemical shift effect in inner electronic levels of Cu due to oxidation, Phys. Rev., 110 (1958) 776.

[13] J.G. Jenkin, R.G.C. Leckey, J. Liesegang, The development of X-ray photoelectron spectroscopy 1900–1960, J. Electron Spectrosc. Relat. Phenom., 12 (1977) 1–35.

[14] S. Hufner, Photoelectron Spectroscopy Principles and Applications, 3rd Edition, Springer, Berlin; Heidelberg, 2003.

[15] J.F. Moulder, W.F. Stickle, P.E. Sobol, K.D. Bomben, Handbook of X-ray Photoelectron Spectroscopy: A Reference Book of Standard Spectra for Identification and Interpretation of XPS Data, Physical Electronics, Eden Prairie, MN, 1995.

[16] T. Deegan, X-ray photoelectron spectrometer calibration and thin film investigations on germanium oxides, Master of Science Thesis, Dublin City University (1998).

[17] S. Tougaard, P. Sigmund, Influence of elastic and inelastic scattering on energy spectra of electrons emitted from solids, Phys. Rev. B, 25 (1982) 4452.

[18] C.J. Powell, A. Jackson, NIST Electron Inelastic-Mean-Free-Path Database, Version 11, NIST, Gaithersburg, MD, 2000.

[19] C.J. Powell, Practical guide for inelastic mean free paths, effective attenuation lengths, mean escape depths, and information depths in X-ray photoelectron spectroscopy, J. Vac. Sci. Technol. A, 38 (2020) 023209.

[20] M.P. Seah, W.A. Dench, Quantitative electron spectroscopy of surfaces: A standard data base for electron inelastic mean free paths in solids, Surf. Interface Anal., 1 (1979) 2–11.

[21] S. Tanuma, C.J. Powell, D.R. Penn, Calculations of electron inelastic mean free paths V data for 14 organic compounds over the 50–2000 eV range, Surf. Interface Anal., 21 (1994) 165–176.

[22] S. Tanuma, C.J. Powel, D.R. Penn, Calculation of electron inelastic mean free paths (IMFPs) VII reliability of the TPP-2M IMFP predictive equation, Surf. Interface Anal., 35 (2003) 268–275.

[23] H.T. Nguyen-Truong, Penn algorithm including damping for calculating the electron inelastic mean free path, J. Phys. Chem. C, 119 (2015) 7883–7887.

[24] M. Hutley, Diffraction Gratings, Academic Press, New York, 1982: pp. 217–221.

2 XPS Analysis and Data Manipulation Tools

2.1. INTRODUCTION TO THE *RXPSG* SOFTWARE AND THE *R* PLATFORM

RxpsG is a software for the analysis of XPS spectra, developed using the R libraries. *R* is a programming language for statistical computing and graphics. It is an implementation of the *S* programming language and is object oriented and interpreted. Therefore, the code could be directly run without a compiler [1–3]. *R* is supported by the *R* Core Team part of the *R* Foundation for Statistical Computing (see *The R-project* website www.r-project.org/). Originally, R was created by statisticians Ross Ihaka and Robert Gentleman to perform a large variety of statistical functions. *R* is an open-source free programming language within the GNU project [4], a Unix-like collection of free packages utilized for developing operating systems as Linux. The GNU was created by R. Stallman in 1983 with the aim to share software and promote research and is now supported by the Free Software Foundation (GNU Operating System www.gnu.org/gnu/thegnuproject.en.html). *R* base functions are written in C and Fortran while more advanced functions are developed in the *R* language. The *R* core is composed by precompiled executables while all the library functions are interpreted. This allows the user to directly access the code of any of the *R* functions, to see how operations are performed and, if needed, to modify them. *R* has an austere command line interface which renders its use sometimes difficult. For this reason, *R* was provided with the *RStudio* interface, to facilitate the use and the development of software. This has fostered the scientific community to develop and share packages resulting in the creation of a very rich set of libraries for any kind of purposes. This made *R* a very popular platform [5–7] with increasing TIOBE index used to rank the popularity of programming languages and related approval [8]. *R* is an integrated set of software functions for data manipulation and calculation and tools for graphical representation of data. It includes:

- an effective data handling and storage facility;
- a complete list of operators for manipulation of arrays and matrices;
- a large collection of libraries for mathematical and statistical analysis;
- complete packages for plotting the analyzed data and saving the graphical outputs using different file formats;
- a well-consolidated programming language rich in instructions enabling the construction of own software.

Generally, *R* is referred to as an "environment" to characterize it as a fully integrated and coherent system where all objects (functions and data) may be implemented. *R* may be extended adding packages which are made following a well-defined protocol which ensures perfect compatibility and robustness.

DOI: 10.1201/9781003296973-2

Among the graphical libraries available of the *R* environment, the TclTk, the Gwidgets2, and Gwidgets2TclTk allow the development of graphical user interfaces (GUIs) rendering the use of software very easy.

RxpsG may be considered as a collection of GUIs developed to perform the spectral analysis of XPS data. *RxpsG* consists of a complete list of tools for reading spectra, making the background subtraction and the peak fitting using appropriate functions, the chemical speciation, and the elemental quantification. Graphical representation of the analyzed spectra can be performed by a complete list of options offering an easy tool to plot data personalizing the graphical output. Also reporting is easy, *RxpsG* provides a facility to summarize the peak fitting results together with the elemental quantification. Both reports and graphical outputs can be easily imported in textual documents (such as Word™ or LibreOffice Writer™) for documentation and for writing manuscripts. *RxpsG* is an open-source software, since it runs in the *R* environment. This has two main advantages: (i) it allows the user to extend *RxpsG* functionalities integrating his own macro and (ii) *RxpsG* is an open project offering to the scientific community the possibility of collaborating to improve/extend the software potentialities. In the following, we will enter into a more detailed description of the software, explaining how it can be installed and used.

2.2. INSTALLATION OF THE SOFTWARE AND GENERAL DESCRIPTION

The installation of the *RxpsG* software is done using *RStudio*. The first step is then to install *R* and the *RStudio* interface on your computer. A complete description of the *R* installation is given in the *R-Project* website www.r-project.org/. In this website are provided the *R* libraries for any kind of operating system including a wide variety of UNIX platforms, Windows and MacOS. To download *R*, the user has to select first the preferred CRAN mirror among the network of servers around the world. These are used as repository to store identical, up-to-date *R* versions and the related documentation. In the *R-Project* website are given also the base information concerning the system requirements needed for the installation.

2.2.1. HARDWARE

For all the operating systems, it is supposed to have administrative privileges needed for the installation of new packages and to have a good network connection to download the *R* and *RStudio desktop*. 256 MB disk space is needed for the *R* installation; another 256 MB is needed for *RStudio desktop*. One should also consider that a little more disk space will be used for the *RxpsG* package and the list of libraries required by this software. As for the memory, it depends on the amount of data to be processed. Data are temporarily stored in the RAM to make them available for the computations. Approximately, 500 MB of disk space is needed to run the *RxpsG* software using *RStudio* interface, since spectral analysis generally does not require the manipulation of huge amounts of data.

2.2.2. INSTALLATION

This section will describe in detail the *R* and *RStudio* installation. On the CRAN website https://cloud.r-project.org/, select the *R* version for your own operating system and then read carefully the instructions. For Windows™ operating systems, installation consists in downloading and running the *R* installer that is as usually done for common programs. In the case of Linux systems, open a terminal and use the following commands:

```
sudo apt-get update
sudo apt-get install r-base-dev
```

In the case of the MacOS, just go in the CRAN-downloads for MacOS and then click on the *R*-x.y.z.pkg (x.y.z stands for the actual *R* version) which will start the installation procedure.

Once *R* is installed, the second task regards *RStudio desktop* which will be used as the interactive console for *RxpsG*. To download the *RStudio* free package, go to the website www.rstudio.com/ and click **Download Free Desktop IDE** (Integrated Development Environment), and then select the *RStudio desktop* compatible with your operating system and follow the instructions to install the software.

Now a good idea is to test if *R* and *RStudio* are correctly installed. Run *RStudio* and in the console type:

```
version
```

You should obtain something similar to:

```
platform        x86_64-pc-linux-gnu
arch            x86_64
os              linux-gnu
system          x86_64, linux-gnu
status
major           4
minor           1.2
year            2021
month           11
day             01
svn rev         81115
language        R
version.string  R version 4.1.2 (2021-11-01)
nickname        Bird Hippie
```

This is what I obtained on an Ubuntu Linux system with installed the *R*-4.12 version. A similar list of information should be obtained also with the other operating systems.

2.2.3.   *RSTUDIO* CONFIGURATION

By default, *RStudio* interface is divided into four parts as illustrated in Figure 2.1, which can be personalized.

On my PC, the top-left window is dedicated to **Source** codes, the bottom-left panel to the **Console**, the top-right panel to **Environment, History, Plots**, and **Connections,** and the bottom-right window to the, **Files, Packages, Help, and Viewer**. You can extend or minimize each of these windows, I suggest having a reasonably wide Console and Package windows. The *RStudio* layout can be configured selecting (see Figure 2.2)

```
MENU → TOOLS → GLOBAL OPTIONS
```

In the *Pane Layout* option, you can select your own *RStudio* appearance.

2.2.4.   *RXPSG* INSTALLATION

The third step concerns the installation of *RxpsG*. This software is still not part of the CRAN libraries. *RxpsG* can be freely downloaded from the GitHub web site: https://github.com/GSperanza/Tar.Gz/. Download the latest *RxpsG-xyz.tar.gz* version, the Downloads folder on your PC can be used. The *RxpsG* package "depends" on a certain number of *R* libraries meaning that it cannot work without those libraries. When installing any package, *R* controls first if there are dependencies. If this is the case, all the required libraries are initially installed and then *R* proceeds to the installation of the package. This also occurs for *RxpsG*, which requires all the dependencies to be satisfied before its installation.

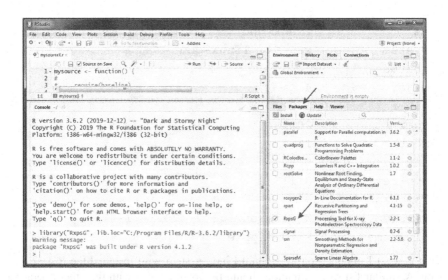

FIGURE 2.1   *RStudio* interface is composed of four windows which in this configuration are **Source, Console, Environment,** and **Packages**.

Under *Rstudio*, copy and paste the following command to install the required libraries:

```
install.packages(c("digest", "gWidgets2", "gWidgets2t-
cltk", "import", "latticeExtra", "memoise", "minpack.
lm", "signal"), repos = "https://cloud.r-project.org",
dependencies=TRUE)
```

Control that installation and proceed correctly without errors. To install *RxpsG*, copy and paste the following command:

```
install.packages(". . . Path-To-Tar.Gz/RxpsG_2.3-2.tar.
gz", type = "source", dependencies=TRUE)
```

where . . .*Path-To-Tar.Gz* is the path to the dowloaded RxpsG_2.3-2.tar.gz file. For windows systems should be something like *C:\Users\Your_User_Name\Dowloads* while for linux systems could be something like *~/Downloads/* supposed the RxpsG_2.3-2.tar.gz was placed in the /Dowloads folder.

Alternatively, you can use the *RStudio* tools to install the libraries. In this case, open the

```
MENU → TOOLS → INSTALL PACKAGES → INSTALL FROM Reposi-
tory (CRAN)
```

and start typing the name of the library to install ("digest", "gWidgets2", "gWidgets2tcltk", "import", "latticeExtra", "memoise", "minpack.lm", "signal"). Select the desired library and press ENTER.

Installation of all the libraries will take some minutes. If *RStudio* is unable to find the needed libraries, first check your network connection, then control which CRAN is set. Open

```
MENU → TOOLS → GLOBAL OPTIONS
```

(see Figure 2.2). This time, click on the Packages icon to switch to the Package options and control if the CRAN mirror corresponds to the default *Global (CDN)— RStudio*. If this mirror is not responding, select the one nearer to your internet connection site and try again.

When all the required packages are correctly installed, open again

```
MENU → TOOLS → INSTALL PACKAGES → INSTALL FROM Package
                                  Archive (.zip .tar.gz)
```

paying attention to select "install from Package Archive" and browse the *RxpsG-xyz. tar.gz* (see Figure 2.3).

Now load the *RxpsG* in the *R* environment. Just type

```
library(RxpsG)
```

this command renders all the package functions available for execution. Alternatively it is possible to load the *RxpsG* library selecting **Packages** in the bottom-right window of the *RStudio* interface (see Figure 2.1). All the libraries installed in your computer will be visualized. Select the *RxpsG* library to load the software.

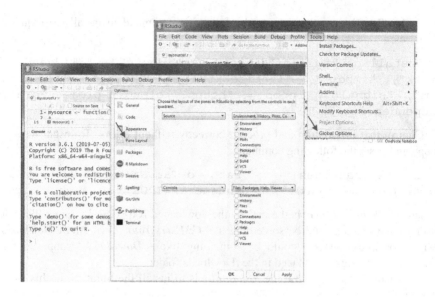

FIGURE 2.2   Selection of the *Pane Layout* in the *RStudio* Global Options.

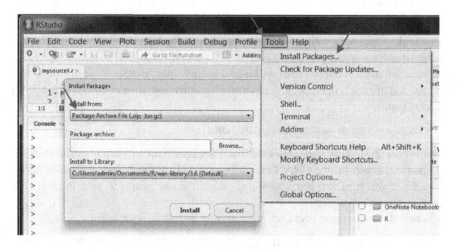

FIGURE 2.3   Package installation from Package Archive **.zip** or **.tar.gz**.

To start working with *RxpsG*, just type

```
xps()
```

This will switch the *RxpsG* interface on, and you can now start the analysis of the XPS spectra. If there is a mismatch between installed R and the compiled version of *RxpsG* a warning message is raised see Figure 2.1. Do not mind this is not a problem.

2.2.5. *RxpsG* Interface and Data Structure

RxpsG is designed to make the XPS analysis easy and intuitive. The *RxpsG* graphical user interface (GUI) is very compact and presents a general menu composed by four main items, *File, Analysis, Plot*, and *Info_Help*, as illustrated in Figure 2.4.

Under menu *File* are the functions used to load data from files of different format. Whatever the file format, the data are loaded into a well-defined structure which is organized for containing all the original information and those deriving from the spectral analysis.

This structure is composed by a *list*, and by a number of *strings*. In *R*, a *list* is a very flexible object which can be used to store data of different nature. The list can contain *strings, numbers, vectors, matrices*, and other lists.

The structure used to store the loaded data is shown here.

```
Formal class "XPSSample" [package ".GlobalEnv"] with
seven slots
```

The symbol "@" indicates a slot in a given structure. As shown, the @.Data slot is a *list()* while the other slots are *chr* indicating that they are of type character. The generic XPS dataset is indicated as *XPS Sample* and is an object of class *XPS Sample*. Any object of class XPS Sample will possess the indicated structure and properties. An XPS acquisition session generally produces a series of spectra. When spectra are loaded into *RxpsG*, they will be placed in the @.*Data* slot. Information about the experimental session will be stored in the @*Project*. In the @*Sample* slot are stored data regarding the analyzed sample, while information about the experiment is saved in @*Comments*. @*User* contains the name of the XPS-operator while the data-file location is stored in the slot @*Filename*. Finally,

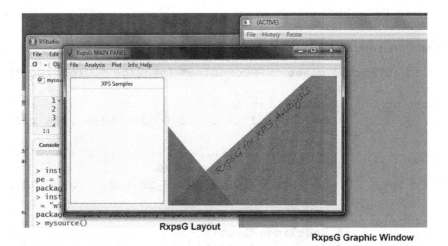

FIGURE 2.4 *RxpsG* layout: the main panel and the graphic window.

the names of all the spectra contained in the data file are listed in *@names*. All these spectra must be stored when you load the XPS data file. Consequently, the *@.Data* slot will be expanded and a number of sub-lists are created corresponding to the number of spectra contained in the XPS-acquisition. It is useful to define another object, the *Core-Line*, so that the XPS Sample may be thought as a collection of Core-Line objects. For each of the spectra, a Core-Line structure is created containing a number of slots namely

@.Data	Raw X, Y spectral data and analyzer transfer function
@RegionToFit	Portion of the spectrum to be fitted. X, Y data
@Baseline	Baseline X, Y data
@Components	Slot expanded to describe All Fit Component
@Fit	Information relative to the Best Fit
@Boundaries	X, Y boundaries of the Region To Fit
@RSF	Sensitivity factor of the Core-Line
@Shift	Energy shift used for energy scale correction
@units	Binding/Kinetic Energy (eV), Intensity (cps)
@Flags	Flags for units, instrument used
@Info	Acquisition information
@Symbol	Core-Line name

An example of XPS data file composed by a Survey, C 1s, and O 1s core-lines loaded into the XPS Sample structure is shown here.

```
Formal class "XPSSample" [package ".GlobalEnv"] with
seven slots

..@. Data:List of 3
```

First Core-Line: Survey

```
.... $:Formal class "XPSCoreLine" [package ".GlobalEnv"]
with 12 slots
.. .. .... @. Data:List of 3
.. .. .. .. ..$: num [1:2611] 1300 1300 1299 1298 1298
                          . . .
.. .. .. .. ..$: num [1:2611]  95194 94634 93975 94176
                          94294 . . .
.. .. .. .. ..$: num [1:2611] 28.6 28.6 28.6 28.6 28.6
                          . . .
.. .. .... @ RegionToFit: list ()
.. .. .... @ Baseline: list ()
.. .. .... @ Components: list ()
```

```
.. .. .... @ Fit: list()
.. .. .... @ Boundaries: list()
.. .. .... @ RSF: num 0
.. .. .... @ Shift: num 0
.. .. .... @ units: chr [1:2] "Binding Energy [eV]"
                                 "Intensity [cps]"
.. .. .... @ Flags: logi [1:3] TRUE TRUE FALSE
.. .. .... @ Info: chr [1:3] "XPS Spectrum Lens Mode:Hy-
brid Resolution:Pass energy 160 Iris(Aper):slot(Slot)"
"Acqn. Time(s): 2611 Sweeps: 1 Anode:Mono(Al (Mono))
(150 W) Step(meV): 500.0 ""Dwell Time(ms): 1000 Charge
Neutraliser:On Acquired On:18/06/27 16:28:13"
.. .. .... @ Symbol: chr "Survey"
```

Second Core-Line: C1s

```
.... $:Formal class "XPSCoreLine" [package ".GlobalEnv"]
with 12 slots
.. .. .... @. Data:List of 3
.. .. .. .. ..$: num [1:301] 293 293 293 293 293 . . .
.. .. .. .. ..$: num [1:301] 1167 1106 1084 1023 1024
                                  . . .
.. .. .. .. ..$: num [1:301] 0.837 0.838 0.838 0.838
                                 0.838 . . .
.. .. .... @ RegionToFit: list()
.. .. .... @ Baseline: list()
.. .. .... @ Components: list()
.. .. .... @ Fit: list()
.. .. .... @ Boundaries: list()
.. .. .... @ RSF: num 0
.. .. .... @ Shift: num 0
.. .. .... @ units: chr [1:2] "Binding Energy [eV]"
"Intensity [cps]"
.. .. .... @ Flags: logi [1:3] TRUE TRUE FALSE
.. .. .... @ Info: chr [1:3] "XPS Spectrum Lens Mode:Hy-
brid Resolution:Pass energy 20 Iris(Aper):slot(Slot)"
"Acqn. Time(s): 602 Sweeps: 2 Anode:Mono(Al (Mono))(150
W) Step(meV): 50.0 ""Dwell Time(ms): 1000 Charge Neu-
traliser:On Acquired On:18/06/27 17:13:37"
.. .. .... @ Symbol: chr "C1s"
```

Third Core-Line: O1s

```
.... $:Formal class "XPSCoreLine" [package ".GlobalEnv"]
with 12 slots
```

```
.. .. .... @. Data:List of 3
.. .. .. .. ..$: num [1:221] 536 536 536 536 536 . . .
.. .. .. .. ..$: num [1:221] 1824 1834 1864 1879 1856
                                  . . .
.. .. .. .. ..$: num [1:221] 0.819 0.819 0.819 0.819 0.819
                                  . . .
.. .. .... @ RegionToFit: list()
.. .. .... @ Baseline: list()
.. .. .... @ Components: list()
.. .. .... @ Fit: list()
.. .. .... @ Boundaries: list()
.. .. .... @ RSF: num 0
.. .. .... @ Shift: num 0
.. .. .... @ units: chr [1:2] "Binding Energy [eV]"
"Intensity [cps]"
.. .. .... @ Flags: logi [1:3] TRUE TRUE FALSE
.. .. .... @ Info: chr [1:3] "XPS Spectrum Lens Mode:Hybrid
Resolution:Pass    energy    20    Iris(Aper):slot(Slot)"
"Acqn. Time(s): 442 Sweeps: 2 Anode:Mono(Al (Mono))
(150 W) Step(meV): 50.0" "Dwell Time(ms): 1000 Charge
Neutraliser:On Acquired On:18/06/27 17:13:37"
.. .. .... @ Symbol: chr "O1s"
..@ Project: chr "RxpsG"
..@ Sample: chr "Z:/X/LAVORI/R/Analysis/Test.vms"
..@ Comments: chr "XPS test data"
..@ User: chr "G.S."
..@ Filename: chr "Test.vms"
..@ names: chr [1:4] "Survey" "C1s" "O1s"
```

Upon spectral analysis, the slots relative to @RegionToFit, @Baseline, @Components, @Fit, and @RSF will be completed with the relative data. In the following, we will refer to *XPS Sample* to indicate a collection of XPS spectra from a certain XPS analysis.

The example above shows how defined objects can be used. Objects are also helpful to construct blocks for larger programs. As seen, an object is a structure possessing some properties or *attributes*. For an object, it is possible to define some methods which, depending on the attributes, will act on the data. Object attributes allow a classification. Classes are used as an outline for the object and contain their definition, their properties, and the methods used to perform specific functions. As an example, in *RxpsG*, a plot function is defined for the XPS Sample and for the Core-Line objects. When plot() is used, *R* controls the kind of data to be plotted. Then plot() works differently if applied to a generic set of data or to an XPS Sample or a Core-Line object, since they have specific attributes and rules. A more precise description of the Object Oriented Programming in *R* is out of the scope of the present book. Detailed information on Object Oriented Programming can be found in [9–11].

2.2.6.   Use of the Memory and .GlobalEnv

Data analysis requires a certain amount of memory to store original data and the results of calculation and manipulation. When data are loaded, R allocates a certain memory in the so-called Global Environment. An environment can be thought of as a collection of objects (functions, variables, etc.). The Global Environment is the top-level environment where the global variables of the main program are saved. If the main program calls a function, a new local environment is created to allow the function to process the data. Variables in the Global Environment are available to any function called by the main program, while the local environment contains only local variables. When the function call ends, the local environment is killed and all the local variables are eliminated. The results obtained applying the function to the data can be made available in the Global Environment by using the return(variable). Another possibility is to use the assign() function which allows for saving the results of a function in a global variable. The definition of local environments renders the program much more stable. If a certain procedure blocks, we can kill the procedure and only the relative local environment and local variables are lost while all the global variables are preserved with the work done up to that moment.

In *RxpsG*, some global variables are defined as the actual XPS Sample, its name, the actual Core-Line, etc. Each of the menu options corresponds to a procedure acting on the XPS Sample or on a given Core-Line. Before exiting the procedure, the results are saved in the global variable using the assign() function. The local environment is killed and generally the results are visualized plotting the modified XPS Sample or Core-Line.

2.3.   BASE OPERATIONS IN *RXPSG*

RxpsG is a software developed for the analysis of XPS Spectra. Essentially, analyzing XPS spectra consists in rather common operations sketched in Figure 2.5.

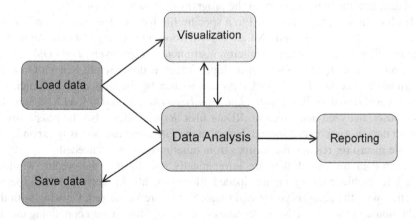

FIGURE 2.5   Logical scheme of the operations commonly made for analyzing data.

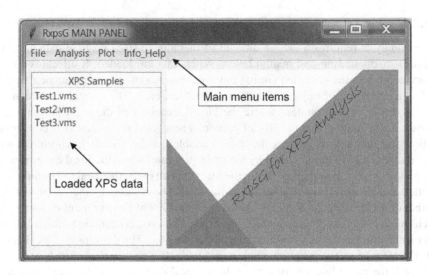

FIGURE 2.6   The *RxpsG* layout.

These operations are organized in four main groups dedicated to *read/save data*, *spectral analysis*, *graphical output*, and *retrieving acquisition information and reporting* and are the four items of the *RxpsG* main menu. *RxpsG* has a rather simple easy and intuitive interface, on top are placed the main menu items, and just below is a window where the names of the loaded XPS data files will appear (see Figure 2.6).

2.3.1. LOAD SPECTRA IN *RxpsG*

File is the first item of the *RxpsG* menu. It contains all the options to load and save data. In the *RxpsG*_2.3-2 version of the software, there are four different options to load data into the main memory of the program as shown in Figure 2.7.

Each of these options is related to a specific file format. The first option is *Load VMS and PXT* data. VMS and PXT are the extensions identifying standard Vamas format and files saved by Scienta-Omicron instruments, respectively. *Load Old Scienta Files* allows reading data saved in multiple folder as done by old Scienta-Omicron instruments. The *Load PXT+RPL data* was written by the authors to load their old Scienta analyzed data files. Finally, *Load Analyzed Data* is used to load XPS Samples previously analyzed and saved in .RData files. *RxpsG* offers also the possibility to import data in *Ascii* textual format. Note that the *File* menu can be easily expanded to include macro for reading file formats from other instruments if needed.

When data are loaded, they are automatically visualized in the graphic window. Just left double clicking on the loaded filenames, all the correspondent spectra together with the analysis performed (peak fitting) are visualized. Clicking with the right mouse button on the XPS filename opens a drop-down menu containing the list of spectra which can be individually selected and visualized (see Figure 2.8).

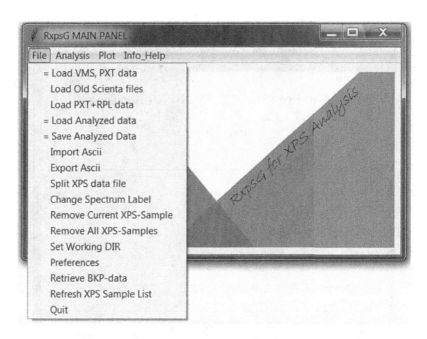

FIGURE 2.7   The options of the *File* Menu item.

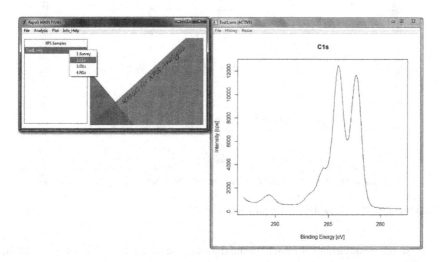

FIGURE 2.8   Double clicking on the XPS filename opens the drop-down list of spectra. Here, the C 1s is selected and visualized.

2.3.2. SAVE ANALYZED DATA

In addition to *Load Analyzed data* reading .RData files with all spectral analysis information, *Save Analyzed data* is used for saving the analyses carried out on the XPS spectra in a file with the extension .RData. The correspondent GUI is shown in Figure 2.9.

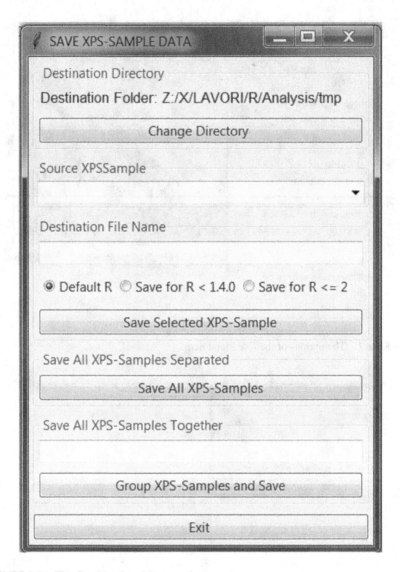

FIGURE 2.9 The *Save Analyzed Data* option panel.

First, the destination folder where the data will be saved has to be set. By default, the folder corresponds to that of the original raw data. If you want to save data in a different folder, just press the *Change Directory* button and browse the new folder as illustrated in Figure 2.10.

Then the *Source file name* corresponding to the XPS Sample to save must be selected. Upon selection, the original parent folder is shown and can be changed as described earlier. Upon selection, also a default name with .RData extension is proposed in *Destination File Name* editable window. The *Destination File Name* can be changed: just type the new *Destination Filename* as indicated in Figure 2.11.

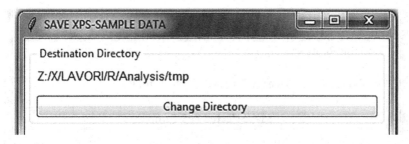

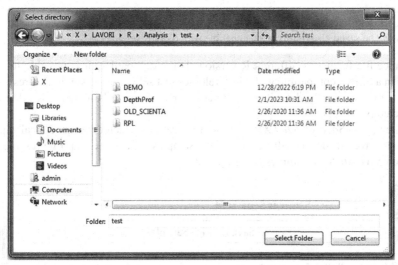

FIGURE 2.10   Selection of the destination folder for saving the analyzed XPS Sample.

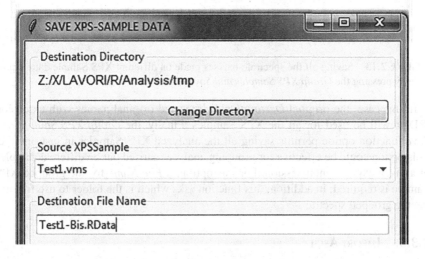

FIGURE 2.11   The original data manipulation Test1.RData made on an XPS Sample can be preserved just using the different Test1-Bis.RData name to save the new spectral analysis.

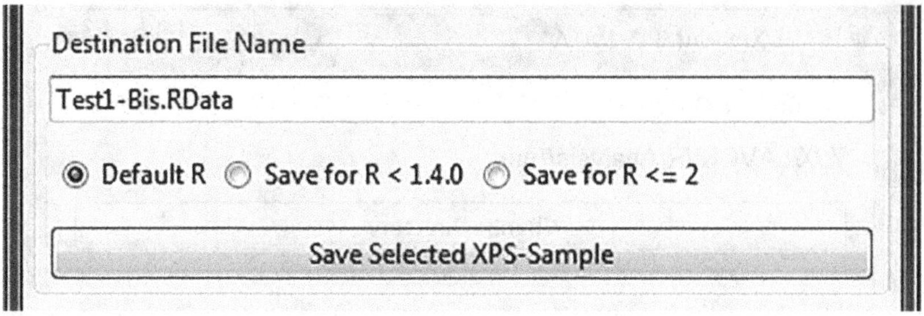

FIGURE 2.12 RData format compatible with the R version installed is selected by default. Files can be saved also in formats compatible with older R versions.

Since file format of different R versions is not compatible, it is possible to save data in a format compatible with old *R* releases just selecting the version correspondent to the own R. By default, file format compatible with the updated R release is selected (see Figure 2.12).

Then press *Save selected XPS Sample* to save the chosen data file. It is also possible to save all the loaded/analyzed XPS Samples in just one step by pressing the button *Save All* XPS Samples (see Figure 2.13).

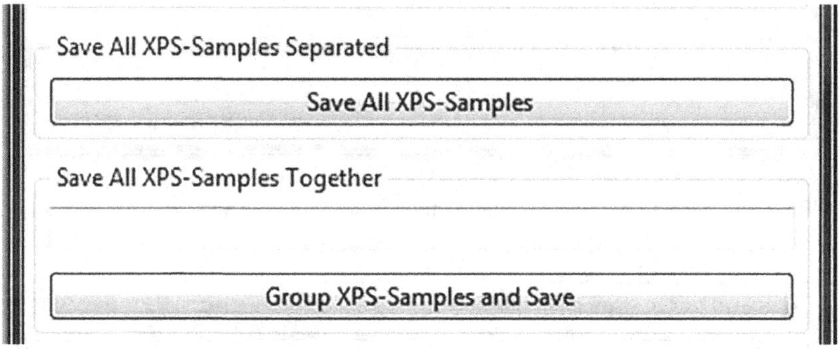

FIGURE 2.13 Saving all the spectral analyses made on different XPS Sample data is possible by pressing the *Group XPS Samples and Save* button

In this case, the original *Destination Folders* and original names with extension .RData will be used for all the XPS Samples. Finally, the *Group XPS Sample and Save* function option permits saving all the analyzed XPS Samples in just one file. This option could be effective for grouping analyses carried out on different samples of a given experimental session. For this option, a *File Name* for the grouped XPS Sample is required; in addition, this function asks which is the folder to use for saving the grouped spectra.

2.3.3. IMPORT ASCII

There are several reasons making the *Import Ascii* option useful. One of them is to plot the results produced by the XPS spectral analysis. Any table of data in text format containing information about the trend of concentrations of a specific element depending on the synthesis process used or on the particular treatment after the synthesis (surface modification via plasma, annealing, chemical treatment, grafting of specific molecules, etc.) can be imported and data plotted using the *RxpsG* functions. Another possibility is to use the *RxpsG* software to analyze spectra from any kind of instrument, provided the possibility to load the data. Generally, Ascii textual format is a common format used to export data from instruments. Then the *Import Ascii* option can be used to load data of diverse nature into *RxpsG* and proceed with the processing. The *Import Ascii* GUI is shown in Figure 2.14.

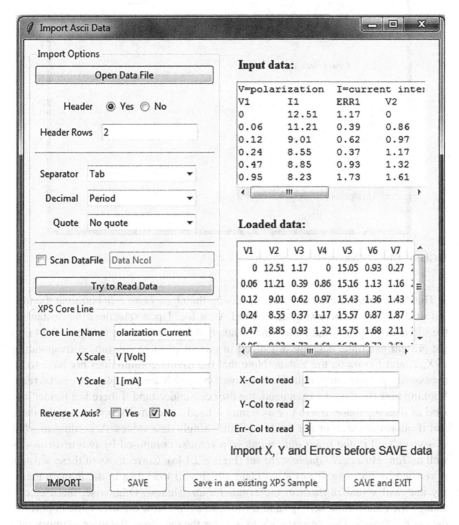

FIGURE 2.14 *Import Ascii* GUI, a text file was selected and its content is visualized in the Input data window. A two-row header was set as well as the data format (*Separator, Decimal,* and *Quote*) in agreement with data structure is shown in the Input window.

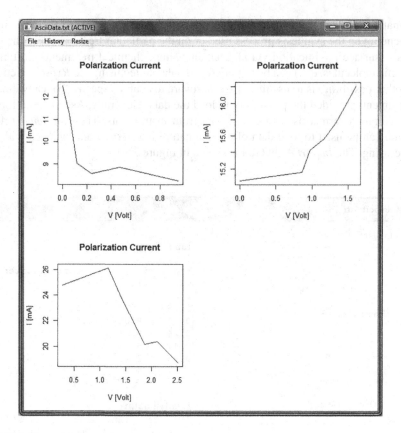

FIGURE 2.14   (Continued)

The Ascii data file can be selected pressing the *Open Data File* button at the top-left of the GUI and browsing the desired Ascii file. Upon selection, the content of the file is displayed in the *Input data* window. As it can be seen in Figure 2.14a, in the present example, data are organized in groups of three columns corresponding to *X*, *Y*, and *Errors* on the Y data. Note that the *Errors* column does not have to be necessarily present and a data file composed by just X and Y columns can be read. Looking how the data are organized, the user can understand if there is a header, the kind of data separator, if a dot or a comma is used as decimal separator for the data, and if quotes are used or not used to identify single data values. According to what was visualized in the *Input data* window, a *Header* composed by two text-rows as well as *Tab*, *Period*, *No quote* were set (Figure 2.14a). Correctness of these settings is verified by pressing *Try to Read Data* which will load data in the R main memory and will visualize them in the *Loaded data* window. Otherwise, the user can change the data-format settings and try to read again. Now the user must import the data in the *RxpsG*. This corresponds to transfer the data from the main memory into the XPS Sample structure. This is done by entering a name for each group of data, specifying the X and Y names and units and if the X-scale has or not to be reversed. In the example of Figure 2.14a, the name of the first group of data is *Polarization*

Current, the X-scale corresponds to *V [Volts]*, while the Y-scale corresponds to *I [mA]*. On the right side of the GUI, are the *X*, *Y*, and *Error column to Read* which in this case is present. Looking at the *Loaded Data* table, the user must enter the number corresponding to the *X*, *Y*, and *Error* columns. For the first group of data, they are columns 1, 2, and 3, respectively (see Figure 2.14a). Then pressing *Import*, the data are loaded into the XPS Sample structure and visualized as illustrated in Figure 2.14b. The user can proceed importing the second groups of data selecting columns 4 and 5 and the additional column 6 for the errors. It is possible to modify the X and Y units and the core-line name, and then data are imported pressing the button *Import* again. Same operations are done to store the third group of data in the XPS Sample structure. The result is shown in Figure 2.14b. *Error bars* are visualized using the *Custom Plot* option where the user can set the more suitable bar style

2.3.4. EXPORT ASCII

Correspondent to *Import Ascii* is the possibility to export data in Ascii textual format using *Export Ascii* shown in Figure 2.15. Exporting data in Ascii format is very easy.

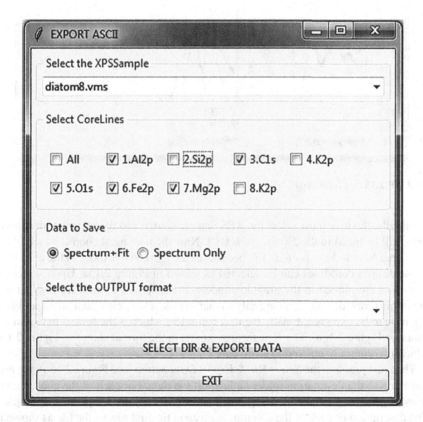

FIGURE 2.15   (a) The *Export Ascii* GUI. In the middle of the GUI, the user can select which of the XPS Sample core-lines will be exported. (b) Each of the selected core-lines is plotted in the graphic window.

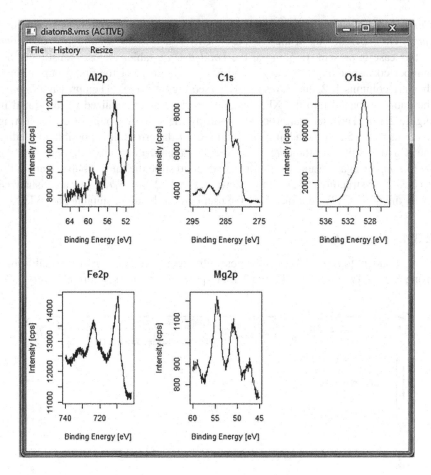

FIGURE 2.15   (Continued)

Initially, the user must select the XPS Sample source and the correspondent core-lines will be listed in the *Export Ascii* GUI. Now the user must choose which of the core-lines have to be exported. *All* checkboxes are marked to export all the core-lines or the desired core-lines can be selected as shown in Figure 2.15a. Upon selection, core-lines are plotted in the graphic window. It is required to indicate whether just the original raw data or also the analysis, namely *Baseline*, *Fit Components* and *Best Fit*, have to be exported. Finally the user can select which is the format to export the data (see Figure 2.16a). The *Raw* format exports in the format shown in Figure 2.16b which is commonly utilized to share data.

Then just pressing the *Select Dir & Export Data* button (see Figure 2.15a), the user can boriwse the destination folder and export the data. For each of the chosen core-lines, is generated a file containing data organized in columns separated by a space. The description of each of the columns is given in the first row of the file as shown in Figure 2.16b. Data in this format are ready to be loaded in electronic worksheets just skipping the first header row.

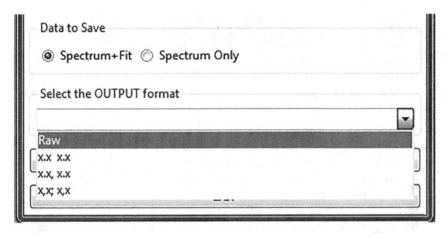

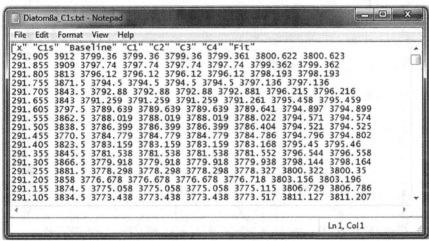

FIGURE 2.16 (a) The possible formats to export data in Ascii format. (b) The Ascii data file generated by the selection of the *Raw* format. The first row contains the description of the columns. Next rows contain data separated by a space character. A *period* is used to separate the decimal numbers.

2.3.5. CHANGE SPECTRUM NAME

It may happen that the user needs to change the name of the XPS Sample or of one or more core-lines. The names of the data file and *Core-Lines* associated to a given XPS Sample are stored among the information relative the acquisition session. The option *Change Spectrum Name* shown in Figure 2.17 allows changing these names.

Upon selection of the XPS Sample, the relative *Core-Lines* will appear in an editable table. Each of the *Core-Line* names can be modified as shown in Figure 2.17 for the O 1s. Also the XPS Sample name in the correspondent editable window can be changed. Pressing the *Save* or *Save & Exit* buttons, all the changes are stored in the main *RxpsG* memory.

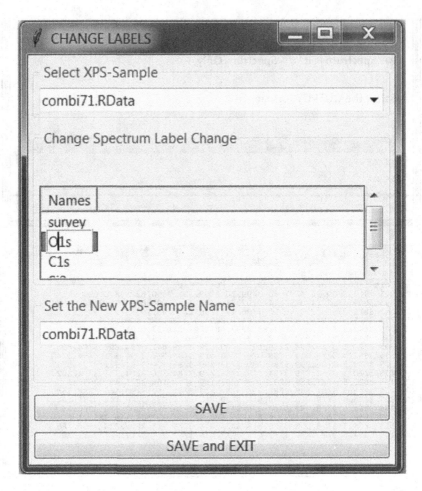

FIGURE 2.17   The *Change Spectrum Label* GUI.

2.3.6. REMOVE XPS SAMPLES FROM MEMORY

When the analysis on a certain XPS Sample or on a series of XPS Samples is accomplished, it is possible to release those objects using the options *Remove current XPS Sample* and *Remove All* XPS Samples. These options delete the actual XPS Sample or all the loaded XPS Samples from the Global Environment. Before using these options, make sure to save the data because otherwise all the work will be lost.

2.3.7. SET WORKING DIR

This option is used to define the working directory where the original XPS data files are located. *RxpsG* will use this directory to load/save the data. The working directory may be changed when necessary. A useful suggestion is to create a folder, let say *XPS_data*, and define this as the default working directory. *XPS_data* will contain a list of subfolders, one for each of the research works. When you load an *XPS Sample*, by default *RxpsG* will open the *XPS_data* folder and you can easily browse your experiment and the desired *XPS Sample*. This renders the work easy and all data well ordered.

2.3.8. PREFERENCES

This option being not directly correlated to the manipulation of spectral data is described in the Chapter 6.

2.3.9. RETRIEVE BACKUP DATA

Sometimes it could happen that the software crashes. Anytime you conclude an operation exiting the correspondent GUI by pressing the *Save and Exit* button, a copy of the work done up to that moment is created. You can restore the work done by loading this backup copy using the option *Retrieve Backup Data*.

2.3.10. REFRESH XPS SAMPLE LIST

This option just updates the list of loaded XPS Samples in the *RxpsG* main panel.

2.4. BUILDING THE *RXPSG* PACKAGE: R OXYGENIZE AND R CMD COMMANDS

As observed previously, *RxpsG* is open source. This allows the user to modify the code of any of the *RxpsG* options or add new macros according to his needs. First, it is necessary to download the whole *RxpsG* package from the GitHub website https://github.com/GSperanza.

The GitHub *RxpsG* repository contains the complete and updated version of the software. The package has a well-defined structure (subfolders and description files) which should not be modified [12]. Enter the *c/your_R_folder/libraries/RxpsG/R* folder and open the desired code with a generic editor and modify it or add a piece of new code. Consider that *RStudio* interface offers a list of options to edit and debug the software. To check if there are errors in the new/modified macro, enter in *RStudio* and just open

```
MENU → CODE → SOURCE FILE . . .
```

and browse your code. Sourcing a code activates the interpretation and the generation of messages in presence of errors. In *RStudio*, there are also options for debugging the program to find errors, where the code is not behaving as expected. To learn about how to use the *RStudio* debugging options, please refer to the official *RStudio* website [13].

Once the code works properly, it is possible to compile *RxpsG* including the changes done. In *RStudio*, you can create a new project selecting the *RxpsG_2.3-2* folder. Just click on

```
PROJECT → NEW PROJECT
```

on top-right of the *RStudio* GUI. This opens the *New Project* window shown in Figure 2.18

There are two options: *New Directory* if the whole package has to be created and *Existing Directory* if, as in our case, the package is already present. Click this option and browse the modified version of the *RxpsG_2.3-2*. Then press the button *Create a Project* and *RStudio* will generate a new *RxpsG* project. In the top-right panel of *Rstudio* will be generated a new section titled **Build**. Here you will have options to build the new *RxpsG* package and check if there are errors.

Another option for Linux systems, is to open the Command Prompt window, change directory, and select the folder containing the new *RxpsG_2.3-2* package. First, the *RxpsG_2.3-2.tar.gz* package must be built. Then type:

```
R CMD build RxpsG_2.3-2
```

This generates the *RxpsG*_2.3–2.tar.gz package. Finally run the command

```
R CMD check RxpsG_2.3-2.tar.gz
```

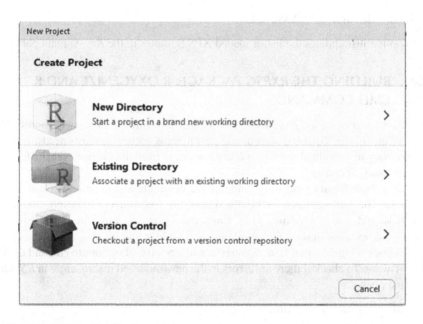

FIGURE 2.18   The New Project window.

The check command makes a rich list of controls on variables of various functions if there are conflicts and if the package can be installed without problems. If compilation is successful, use the *RStudio* option to install the new package as previously described. A complete documentation explaining the construction of an R-package is described in the R website [14].

REFERENCES

[1] J.H. Maindonald, Using R for Data Analysis and Graphics Introduction, Code and Commentary, Centre for Mathematics and Its Applications-Australian National University, 2008. https://cloud.r-project.org/doc/contrib/usingR.pdf.

[2] R.D. Peng, R Programming for Data Science, Lulu.com, https://bookdown.org/rdpeng/rprogdatascience/, 2016.

[3] F. Morandat, B. Hill, L. Osvald, J. Vitek, Evaluating the design of the R language, in: "Object-Oriented Programming", European Conference on Object-Oriented Programming, vol. 7313, Springer, Berlin, Heidelberg, 2012. doi: https://doi.org/10.1007/978-3-642-31057-7_6.

[4] F.P. Deek, J.A.M. McHugh, 8 – The GNU Project, in: Open Source Technology and Policy, Cambridge University Press, Cambridge, 2007: pp. 297–308.

[5] J. Lai, C.J. Lortie, R.A. Muenchen, J. Yang, K. Ma, Evaluating the popularity of R in ecology, Ecosphere, 10 (2019) e02567.

[6] B.K. Hackenberger, R software: Unfriendly but probably the best, Croat. Med. J., 61 (2020) 66–68.

[7] S. Cass, Top programming languages 2021, IEEE Spectrum, https://spectrum.ieee.org/top-programming-languages-2021, 2021.

[8] Index for January 2022, TIOBE the Quality Software Company, Eindhoven, 2022. https://www.tiobe.com/tiobe-index/.

[9] H. Wickham, Advanced R, 2nd Edition, Chapman and Hall/CRC, Boca Raton, FL, 2019. doi: https://doi.org/10.1201/9781351201315.

[10] T. Mailund, Advanced Object-Oriented Programming in R: Statistical Programming for Data Science, Analysis and Finance, Apress, Berkeley CA, 2017. doi: https://doi.org/10.1007/978-1-4842-2919-4.

[11] K. Black, R Object-Oriented Programming, Packt Publishing Limited, Birmingham, n.d.

[12] R Core Team, Writing R Extensions, https://cran.r-project.org/doc/manuals/r-release/R-exts.html, 2021.

[13] J. McPherson, Debugging with the RStudio IDE, https://support.rstudio.com/hc/en-us/articles/205612627-Debugging-with-the-RStudio-IDE, 2021.

[14] H. Wickham and J. Bryan, R Packages. Organize, Test, Document, and Share Your Code, 2nd Edition, O'Reilly Media Inc., Sebastopol, CA, 2023. Online version of the 2nd edition of the book: https://r-pkgs.org/.

3 Acquisition of Spectra and Analysis

3.1. ACQUISITION OF SPECTRA

Materials interact with the environment through the surface. The physical and morphological properties and the chemical composition of these surfaces decide the nature of the interactions. For example, the surface morphology together with its composition determine the surface energy [1, 2]. They govern the surface wettability and adhesive properties [3]. The surface chemistry is involved in the corrosion processes [4–6]. Catalytic efficiency is linked to chemical and physical properties like the system electronic structure and its size [7, 8]. The surface composition also affects the contact potential [9, 10], and failure modes are related to the surface properties [11]. Therefore, surfaces and their chemistry are crucial in defining the behavior of a solid toward the surroundings.

XPS is a powerful technique providing data essential to define the physical and chemical properties of the material. The list of information obtained is very rich: the surface composition and the chemical state of the composing elements, their abundances and how these concentrations vary with depth in the first subsurface layers. Finally, XPS also indicates if the surface composition is homogeneous or not. Specific acquisition conditions emphasize one or the other of these aspects. In spectral acquisition mode, high-resolution spectra will show which chemical bonds are formed and which elements are bonded together. The sample inclination enhances the bulk or surface components and how the concentrations change with depth. In imaging modality, it is possible to visualize the distribution of selected elements in the analyzed area. Valence bands describe the electronic structure of the outer orbitals of the material defining the electronic and optical properties.

This chapter is dedicated to describe how to set up the acquisition of spectra and how to perform the spectral analysis for obtaining the aforementioned information.

3.2. SAMPLE PREPARATION AND OPTIMIZATION OF THE ACQUISITION CONDITIONS

XPS analyzers operate in ultra-high vacuum (UHV) requiring samples to be compatible with this working condition. To satisfy this request, analysts must carefully prepare the samples. This issue becomes essential dealing with high-sensitive analytical instruments. In the case of XPS, we should consider the limited sampling depth of the technique which emphasizes the contribution of the surface. It is then immediately clear that presence of contaminations on the sample surface may create serious drawbacks. Then good-practice rules should be adopted to prepare the sample for the analysis. First is suggested to wear glows to avoid skin contacts with the sample surface.

DOI: 10.1201/9781003296973-3

This is beneficial for both maintaining the samples clean and avoiding desorption of organic molecules which can be detrimental for vacuum during insertion under UHV. Also tools and mounting stage must be always cleaned to reduce contaminations and operations should be made in an appropriate room space. Generally, highly volatile highly pure solvents are used to remove contaminants such as hydrocarbons. Samples have to be fixed to the mounting stage using appropriate clips (normally provided by the instrument manufacturer) ensuring a good electrical contact to get the sample at the same reference potential of the instrument. It must be verified that the clips do not interfere with the analyzed area especially in the case when angle-resolved analyses have to be carried out. Vacuum compatible double face graphitic based tape can be used to fix the sample to the carrier. This tape can be useful especially in the case of irregular shaped samples. Testing the degree bonding is important to avoid sample loss during insertion or tilting, since recovery of the specimen would require instrument venting. Analysis of powders requires particular attention during preparation. One solution is to pelletize the powder applying high pressure. If this cannot be done, a possible solution is to place the powder in a cavity of the carrier to avoid losses during insertion/extraction of the specimen from the analysis chamber. Attention must be paid evacuating the load-lock chamber, because if done too quickly, it will disperse the powder in the vacuum system. A widely used method is to place the powder on a soft indium foil or on a double side tape and press it trying to obtain a flat surface. The use of clean indium has the advantage that interference among the powder and indium spectral components is improbable. This could not be the case when adhesive tape is used and organic components are also part of the specimen. Another solution is to press the powder between thin metal foils as aluminum sheets. Then a reasonably small window can be opened in the upper foil letting the powder enclosed in the metal pocket. This helps avoiding the powder dispersion and the charge compensation. Another possible solution is to disperse the powder in an appropriate solvent. Put some drops on a substrate and let them to dry leading to residuals. Because the powder layer is very thin, likely contribution may derive from the substrate which must be carefully selected to avoid spectral interference.

After sample preparation, the sample is placed in a pre-chamber, the load-lock chamber, to evacuate the atmosphere till a vacuum typically in the range of 10^{-7}–10^{-8} mbar. Then the sample carrier is introduced in the analysis chamber kept in a UHV ranging from 10^{-9} to 10^{-10} mbar and mounted on a sample-holder. Best clean surfaces are obtained by scraping or fracturing the sample in UHV to avoid immediate oxidation and contamination. Clearing adventitious carbon may be done by argon sputtering. Argon ion bombardment removes the top layers of the sample surface. In ion sputtering, Ar^+ ions at energy of some KeV hit the surface with a certain inclination leading to erosion of the outmost atoms. A fraction of the Ar^+ ions, however, penetrate the material below the surface. This process induces the formation of defects and, for prolonged sputtering times, amorphization of the sample structure.

After the surface is prepared for the analysis, the sample has to be moved under the analyzer. In all the XPS instruments, there is a well-defined position where the sample has to be placed for the analysis. This position corresponds to the focus of the electrostatic lenses needed to transport electrons to the analyzer (Figure 3.1). In a monochromatized X-ray source, this position corresponds also to the focus of

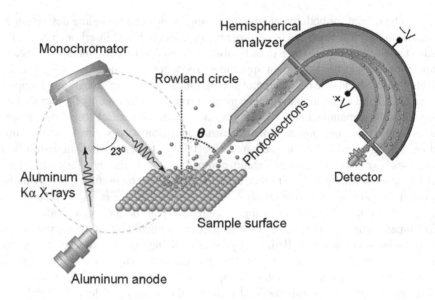

FIGURE 3.1   Schematics of an XPS instrument. The X-source, the monochromator crystals, and the sample are located on the Rowland circle (dashed line). The focused X-ray beam on the sample surface corresponds to the focus of the acceptance electrostatic lens of the analyzer.

the monochromator to maximize the X-ray spot intensity on the sample surface. As described in Section 1.5, XPS systems are designed to locate sample, quartz crystals, and X-ray source on the surface of a Rowland focusing circle (see Figure 3.1).

As observed in Section 1.5, magnetic lenses are used to increase the collection of photoelectrons, thus compensating for the decreased intensity of the X-rays to the sample caused by the monochromator. The typical solid angle of pure electrostatic lenses is ~25⁰ is increased to 90⁰ with the use of magnetic lens resulting in a higher instrument sensitivity. Magnetic lenses also offer a lower aberration coefficient which improves the spatial resolution when the instrument is operated in an imaging mode [12].

3.3.   ENERGY SCALE REFERENCING, CONDUCTING AND NON-CONDUCTING SAMPLES

Optimization of the signal intensity can be done if the sample is conductive. In the case of insulating specimens, photoemitted electrons leave a positively charged surface and the electric field generated can be sufficiently high to block further emission of electrons. In this case, not any signal will be detected. In some cases, the sample displays a limited conductivity leading to a partial compensation of the lost charge. A photoelectron spectrum is generated but distortions are present and its position is shifted at a higher binding energy. These effects are limited when the photoemission is induced by non-monochromatized sources. The absence of the monochromator allows electrons form the source to reach the sample source thus providing a neutralization of the surface charge. In monochromatized instruments, charge is not neutralized and the high photon density generated by the focusing quartz crystals requires charge compensation. This latter is done by irradiating the sample surface with low-energy electrons. To prevent broadening of the XPS peaks, the neutralization must be uniform all over the analyzed area. The optimal compensation conditions are obtained by looking at a core-line single peak trying to make the lineshape the more symmetric the possible while minimizing its full width at half maximum (FWHM). This procedure gets complicated in presence of chemical bonds resulting in additional spectral features requiring a certain experience to recognize optimal compensation conditions. Generally, these conditions lead to an excess of "flooding" electrons with a negative charging of the sample surface. As a consequence, the photoelectron spectra are shifted with respect to their canonical position and a binding energy alignment to the correct position is required during data processing. In particular cases, when highly insulating samples with irregular surface are analyzed, neutralization may be very complex and residual charge remains in spite the attempts to optimize the charge compensation. In these cases, some possible countermeasures may be considered [13]. First, decrease the X-ray photon flux density decreasing the dissipated power at the anode. Less photons result in less photoelectrons generated with lower charging effects. However, this requires longer acquisition times which could be a problem considering that X-rays possess sufficient energy to break chemical bonds [14]. Second, use an earthed conductive grid placed on the sample surface; third, surround the analysis area with a conductive metal foil. This helps the removal of the electrostatic charge thus neutralizing the sample. Last, insulating powders can be pressed onto an indium foil to help the elimination of the charge buildup.

After optimization of the charge compensation and acquisition of the spectra, energy scale alignment is needed to correctly interpret the spectra and assign the chemical bonds. In the case of conducting samples in contact with the spectrometer, spectra are referred to the position of the Fermi level which must correspond to zero binding energy. Analyzers are calibrated by making the Fermi level (E_F) to fall at BE = 0. Calibration is performed acquiring the valence band in reference materials, generally silver or gold, showing a well-distinct Fermi Edge. An example of the silver valence band and the relative Fermi Edge is reported in Figure 3.2.

For non-conducting samples, the position of the Fermi level cannot be defined using the valence band edge. For insulators, the valence band edge does not correspond to the Fermi level, because E_F falls between the valence band and the conduction band in the energy gap. Therefore, for insulators, there is not a reference energy level to calibrate the binding energy scale. In addition, any sample presents a peculiar charging meaning that energy referencing has to be carried out for each specimen. In these circumstances, internal referencing of the BE scale is the best and commonly applied procedure. This is accomplished selecting a particular peak of the spectrum which can be related to a well-defined/accurate value of binding energy [15]. This reference peak is visualized and forced to fall at the correct position. The energy shift is applied also to all the other acquired spectra presuming that they are affected by the same "charging" potential. The peak assigned to adventitious carbon

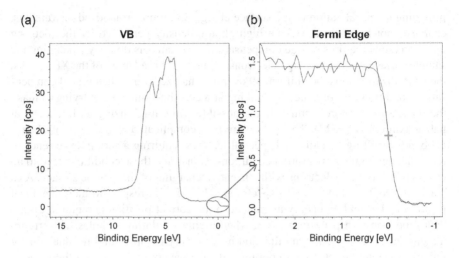

FIGURE 3.2   (a) Valence band of sputter-cleaned Ag. Indicated is the spectral region straddling the Ag Fermi level which is amplified in (b). The Fermi level is indicated by the cross. This value corresponds to BE = 0 eV as expected for calibrated analyzers.

is one of the more frequently utilized references. It is known that the hydrocarbon peak falls at 285 eV [14] and, as a contaminant, is not involved in chemical bonds with the material.

3.4.   *RXPSG*: SELECTION OF A REFERENCE AND ENERGY SCALE ALIGNMENT

Energy scale alignment in *RxpsG* is very easy as described in the following example for a polyetheretherketone (PEEK) specimen. Figure 3.3 shows the *Energy Shift GUI*.

On the top of the GUI, it is possible to select the XPS-Sample and the core-line among the spectra acquired from PEEK sample. As observed in the previous section, the hydrocarbon peak is a good reference and then the C 1s is selected as shown in Figure 3.3c. The fit component at lower BE corresponds to the CH_x and must be located at 285 eV. Upon selection of the component C1, the correspondent binding energy 280.3 eV as obtained from the fit will appear in an editable window where the correct energy may be written as shown in Figure 3.3e. If the core-line fitting is not done, as in the majority of the cases, pressing the *Cursor* button, it is possible to define the CH_x position by hand. To increase the precision, a zoom around the CH_x maximum may be done pressing the *Zoom Corners* button to define two diametrical corners of the zooming area (see Figure 3.3d). On the bottom will appear the actual cursor position which must be forced to 285 eV. In the present case, a positive shift of +4.7 eV is applied to all the spectra. It is possible to apply the shift to the single spectrum by choosing the *Selected core-line only* option (see Figure 3.3c). The button *Save and Exit* will store the aligned spectra of the XPS-Sample in the. *GlobalEnv*.

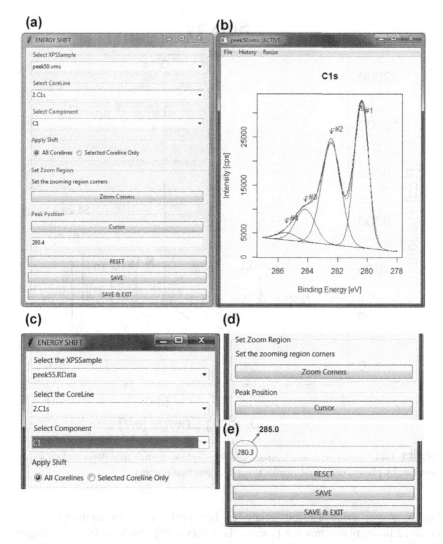

FIGURE 3.3   (a) The *Energy Shift GUI*; (b) the C 1s chosen as reference spectrum for energy alignment; (c) the C1 component of the C 1s fit corresponds to CH$_x$. The energy shift can be applied to all core-lines of just to the C 1s. (d) In the absence of fit, the reference position can be provided using the cursor. A zoom is available to increase the precision. (e) The original and the new BE corresponding to the CH$_x$ carbon component at 285.0 eV.

3.5.   WIDE SPECTRA

The wide spectrum contains the signature of all the elements present in the analyzed materials. Generally XPS spectra are dominated by sharp peaks corresponding to core-level photoemission. An example is shown in Figure 3.4, reporting a wide scan of an amorphous hydrogenated carbon film. The survey spectrum is composed by a few peaks in different BE positions. As observed in Chapter 1 of this book, the

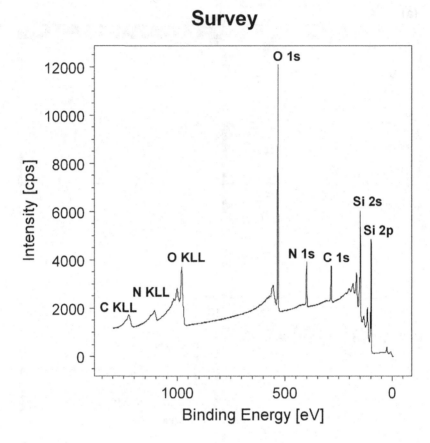

FIGURE 3.4   Survey scan of a hydrogenated amorphous carbon film deposited on a silicon substrate by plasma enhanced chemical vapor deposition process.

binding energy value is strictly related to the emitting atom and to the energy level involved in the photoemission process. This peculiarity can be utilized to recognize the elements present on the analyzed sample surface. In Figure 3.4 are indicated the elements detected on the surface of a silicon substrate deriving from an ultrathin hydrogenated amorphous film deposited using a PECVD system. Silicon contributes with the 2p and 2s core-lines. It is also visible a sequence of oscillations on the high BE side of these core-lines which represent the loss structures due to the excitation of plasmons. At increasing the BE values, follow the peaks corresponding to the 1s orbital of carbon, nitrogen, and oxygen. At higher BE values are located the Auger features of these three elements. These are the elements expected considering that the gaseous precursors used in the PECVD process are CH_4 and NH_3.

Another example is shown in Figure 3.5. Here are shown spectra deriving from four transition metal nitrides namely scandium, titanium vanadium and chromium nitrides [16–19]. As it can be seen at the resolution used to acquire the wide spectra, the N 1s peaks fall at ~400 eV in all the survey spectra. Different are the core-lines

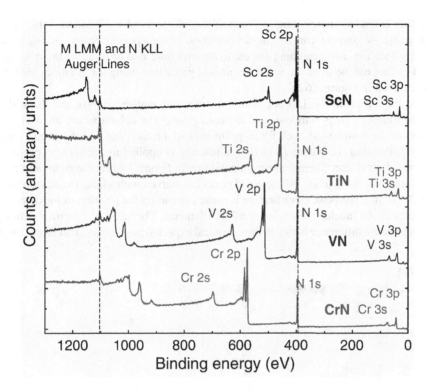

FIGURE 3.5 *In situ* XPS survey spectra of first-row transition metal nitrides ScN, TiN, VN, and CrN.

Source: Adapted with permission from refs. [16–19]

deriving from the 2s and 2p orbitals of the metals. Increasing the atomic number from Sc to Cr, the BEs relative to the 2s and 2p core-lines shift to higher BEs. This situation is replicated in the Auger spectra. Figure 3.5 shows that the Auger LMM components deriving from the transition metals shift to lower BE from Sc to Cr while the position of the peak generated by N at ~1,110 eV remains unchanged through the spectra.

Figure 3.5 clearly shows the specificity of XPS in detecting different chemical elements. Increasing the atomic number by one unit from 21 of Sc to 24 of Cr is sufficient to cause sensible changes in the spectral features allowing unambiguous assignment to a selected element. For this reason, the survey spectra are acquired on a large energy range to map all the elements of the Periodic Table (except hydrogen and helium) and allow the speciation of the analyzed material.

3.5.1. ASSIGNING SPECTRAL FEATURES TO CHEMICAL ELEMENTS IN *RXPSG*: AUTOMATIC PROCEDURE

Assignment of spectral feature to chemical elements means finding the composition of the analyzed material. This operation is performed on survey scans acquired on a rather extended energy range to map all the elements of the Periodic Table. The automatic assignment consists in the detection of the spectral peaks, getting their energy location, and searching for elements matching one or more of the detected peaks. This can be done using an automated procedure using the *Element Identify* GUI shown in Figure 3.6a.

On the top of the GUI, it is possible to select the degree of noise and the width of the running energy window used to detect peaks. The default values are 7 and 5, respectively, which work well for a spectrum with a reasonable signal-to-noise ratio (SNR). Pressing the button *Detection*, a filtering is applied to the survey spectrum (survey dashed line, filtered spectrum solid line in Figure 3.6a on the right) and recognition of peaks is made. All detected peaks are marked with a label (a circle in Figure 3.6 right). The peak identification is made comparing the position of the detected peaks with the tabulated energies of all the elements. The value of *Precision* defines the maximum difference between the canonical expected position and that found from

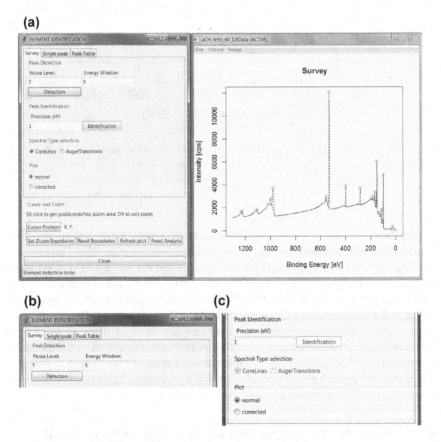

FIGURE 3.6   (a) *Element Identification* GUI on the left and result of peak detection on the right. (b) Noise level and energy window settings used for the peak detection. (c) By default, 1 eV is the precision used to identify the detected peaks by comparison with tabulated data of Core-lines in this Figure.

the real spectrum which is accepted for the assignment of an element to a spectral feature. 1 eV is indicated by default as shown in Figure 3.6c. Consider that the tabulated energies refer to pure elements while chemical shift may affect the same element in the real spectrum. By default, comparison between real and canonical positions is made using the *Core-Line energy Table*. However, it is possible to run the comparison using the *Auger Transition Energy Table* (see Figure 3.6a). Peak identification is activated by pressing the button *Identification*. When peak position is not found among the canonical core-line energies, a warning is displayed. This can occur, for example, when positions of Auger components are compared with core-line energies or when the chemical shift affecting a peak is higher than the value of the Precision given.

Diversely, when an element possesses spectral components falling in correspondence of the position of one detected peak, a different message is hit indicating the element found and showing the correlations with the identified peaks. An example is shown in Figure 3.7 for the Sm element. A correspondence between the position of the core-lines of Sm indicated by vertical bars and those of the identified peaks is found at ~1,100 eV and at ~285 eV. However, this correspondence does not hold for all the other Sm core-lines and Auger transitions.

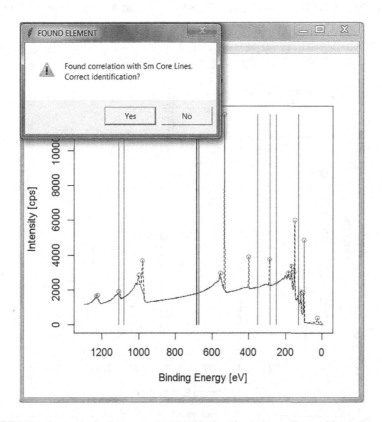

FIGURE 3.7 A correlation between the identified peaks and core-lines of Sm is found only for energies ~ 1,100 eV and ~285 eV. Lighter bars correspond to the Sm core-line positions and bars falling a ~ 680 eV indicate the position of the Auger components.

This indicates that Sm is not an element present in the analyzed sample, and therefore Sm must be discarded answering *NO* to the warning (Figure 3.7). Different is the case of oxygen where perfect correspondence is found for all the spectral components as shown in Figure 3.8. Vertical bars in this case fall at the exact energy of some of the survey peaks. The remaining peaks are generated by different elements (N, C, and Si). In this case, to add oxygen to the list of assigned elements, the user must answer *YES* to the warning. This procedure is repeated for all the elements possessing a spectral feature falling nearby a survey peak with the indicated *Precision*. Spectra can be visualized in *Normal* or *Corrected* modality (Figure 3.6c). Finally, the peak identification and assignment procedures may be repeated by pressing *Reset Analysis*. This allows reinitializing the noise level, the energy window extension, and the search precision used for the element identification. The second page of the *Element identification GUI* is shown in Figure 3.9. This page regards the identification of single peaks.

To start the procedure, manually identify the peak position with the mouse pressing the button *Press to Locate* (Figure 3.9c). The position is marked by a circle (Figure 3.9b). Input the *Precision* used to recognize the element, then select which energy Table to utilize, the one for *Core-lines* or that for *Auger Transitions*, and

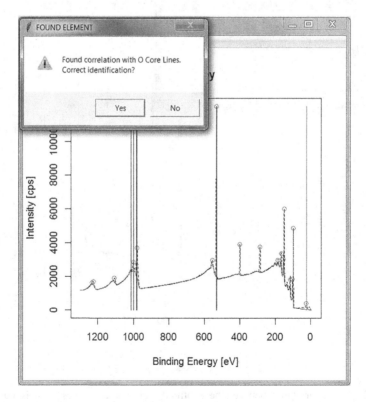

FIGURE 3.8   Oxygen spectral components, perfect agreement is present for 1s and 2s components at ~530 eV and 20 eV respectively and with KLL Auger components at ~ 1,000 eV.

finally select the criterion which will be used to recognize the element shown in Figure 3.9d: (1) the element closest to the *Marker Position*; (2) the element possessing the higher RSF inside the energy range *Marker Position ± Precision*; (3) find all the elements with spectral components in the energy range *Marker Position ± Precision*. It is possible to zoom around the spectral peak pressing the button *Set Zoom Boundaries* and defining the position of the opposite diametrical corners with the mouse (see Figure 3.9e). The button *Reset analysis* allows restarting the procedure of peak identification and assignment.

The last page of the *Element Identification* GUI, the *Peak Table* page, is shown in Figure 3.10a. Selection of *Core-Lines* or *Auger Transitions* will switch between the Tables relative to these spectral features. It is possible to retrieve the position of the

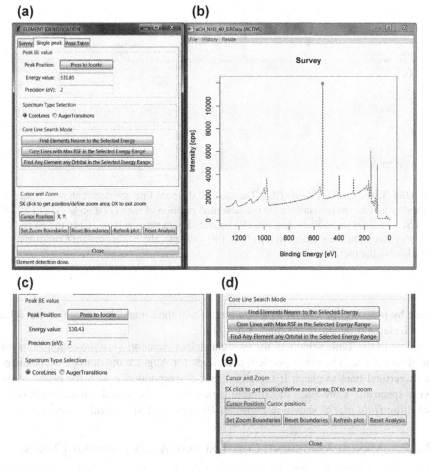

FIGURE 3.9 (a) *Single Peak Identification* GUI. (b) The peak selected for identification is marked with a circle (right panel). (c) Peak can be identified by mouse locator and the selected position is visualized. (d) The criterion which can be used to identify the selected peak. (e) To increase the precision, an expanded view of the peak can be done.

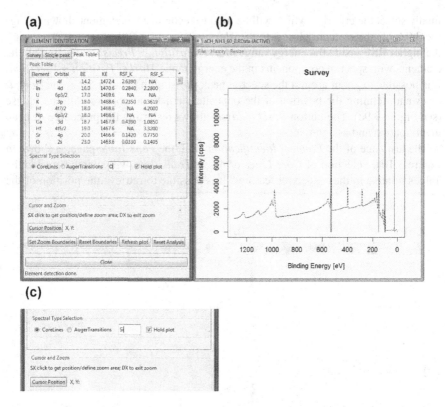

FIGURE 3.10   (a) *Peak Table* GUI. The table of the core-line position is shown. *Hold plot* modality is selected to visualize (b) the tabulated position of both O (at ~530 eV and 20 eV) and Si core-lines (at ~150 eV and 100 eV) indicated with vertical bars. (C) In the entry field the user can input the element to test, Si in this example, the correspondent Core-lines will be added to the plot.

peak by pressing the *Cursor Position* button and then manually search in the Table which element falls near the marker position.

Close to the Table selection, the user can select element, Si in the example shown in Figure 3.10c. The correspondent core-lines (or Auger transitions) are visualized with vertical bars to check if there is or not correspondence with the peaks of the wide spectrum. Core-lines of different elements may be visualized simultaneously selecting *Hold plot* as shown in Figure 3.10b for O and Si spectral components.

3.5.2.   Manual Assignment: Core-Lines and Auger Component Tables

Automatic procedure to identify element spectral components in wide spectra may be boring for users who have a certain experience and immediately recognize the spectral components of the more common elements. In this case, they can utilize the *Core-Line and Auger Transition Tables* option reported in Figure 3.11.

CORE LINE AUGER TRANSITION TABLES

Peak Table

Element	Orbital	BE	KE	RSF_K	RSF_S
Hf	4f	14.2	1472.4	2.6390	NA
In	4d	16.0	1470.6	0.2840	2.2800
U	6p3/2	17.0	1469.6	NA	NA
K	3p	18.0	1468.6	0.2350	0.3619
Hf	4f7/2	18.0	1468.6	NA	4.2000
Np	6p3/2	18.0	1468.6	NA	NA
Ga	3d	18.7	1467.9	0.4390	1.0850
Hf	4f5/2	19.0	1467.6	NA	3.3200
Sr	4p	20.0	1466.6	0.1420	0.7750
O	2s	23.0	1463.6	0.0330	0.1405
Y	4p	24.0	1462.8	0.1410	0.0910
Sn	4d	24.0	1462.6	0.9900	2.7000
Ta	4f7/2	25.0	1461.6	3.0820	4.8200
U	6p1/2	25.0	1461.6	NA	NA
Ca	3p	26.0	1460.6	0.2350	0.5070
Ta	4f5/2	27.0	1459.6	NA	3.8000
Zr	4p	29.0	1457.6	0.2820	1.0500
Sc	3p	29.0	1457.6	0.2280	0.6500
Ge	3d	29.3	1457.3	0.5360	1.4200

Auger Transitions

Element	Transition	BE	KE	RSF_K	RSF_S
Ru	NOO	1453.6	33.0	0.96	0
I	NOO	1450.6	36.0	0.60	0
Rh	NOO	1449.6	37.0	0.43	0
Li	KLL	1448.6	38.0	0.37	0
Pd	NOO	1446.6	40.0	0.43	0
Cs	NOO	1441.6	45.0	0.27	0
Fe	MNN	1441.6	45.0	1.00	0
Ir	OPP	1435.6	51.0	1.45	0
Co	MNN	1435.6	51.0	1.21	0
Ba	NOO	1434.6	52.0	0.30	0
Br	MNN	1433.6	53.0	0.75	0
Ga	MNN	1433.6	53.0	0.32	0
Ba	NOO	1431.6	55.0	0.20	0
Rb	MNN	1431.6	55.0	0.37	0
Cu	MNN	1430.6	56.0	0.85	0
Zn	MNN	1429.6	57.0	1.16	0
Cu	MNN	1427.6	59.0	0.67	0
Ni	MNN	1427.6	59.0	1.02	0
Ce	NOO	1426.6	60.0	0.24	0

aCH_NH3_60_II.RData ▼   Element1   ☐ Core Lines   ☐ Auger Transitions   ☐ Hold plot   **CURSOR POSITION**

UNDO   **REFRESH**   **CLOSE**

FIGURE 3.11   The *Core-Line and Auger Transition Tables* GUI.

Essentially, this GUI replicates what already described for the third notebook page of the *Element Identification* option. The only difference is that here both the *Core-Line* and *Auger Transition Tables* are available. After selection of the *XPS-Sample*, the user can obtain the position of a given peak by pressing the button *Cursor Position*. Then can scroll the *Core-Line* or the *Auger Transition* table to find which are the elements possessing spectral features near the peak position. Specific elements can be indicated using the input field to visualize the position of the correspondent core-lines and Auger components with vertical bars respectively. Also in this GUI, *Hold plot* allows showing the position of spectral components from different elements. The button *UNDO* is pressed to cancel the last bars added to the wide spectrum if no correspondence occurs. The *REFRESH* button cleans all the bars from the survey spectrum.

3.5.3. EXTRACT SPECTRAL FEATURES FROM WIDE SPECTRA

Another option involving the Survey is the *Extract from Survey* used for clipping a portion of the wide spectrum. The GUI is shown in Figure 3.12a. In the upper part of the GUI, it is possible to select the XPS Sample and the spectrum to process. By default, the Survey spectrum of the active XPS Sample is shown as depicted in Figure 3.12b. However, the extract GUI works on all the kind of spectra namely wide spectra, core-lines, Auger features, and valence bands.

When the Extract GUI is opened, the mouse is active to define the area to clip as shown in Figure 3.12 on the right. The button *Select Region* extracts the portion of the spectrum defined by the frame. The plot is restricted to the selected area which can be refined moving the four crosses at the corners in the desired position to optimize extension of the energy range (Figure 3.12c). This is easily done just left-clicking near the corner in the new position and the frame will adapt correspondingly. This is illustrated for the left corners in the image.

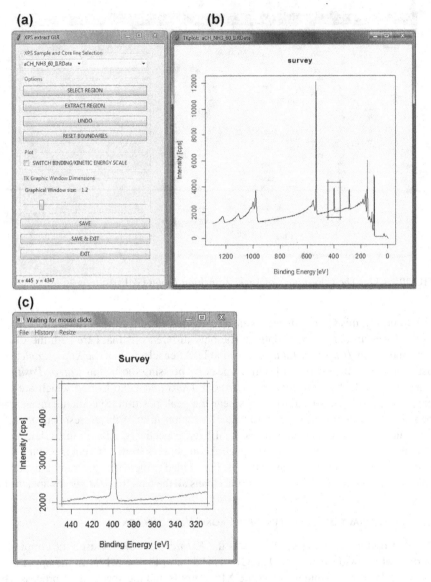

FIGURE 3.12   *Extract from Survey* GUI (left) and the identification of the region to clip (right).

UNDO or *RESET BOUNDARIES* can be used to restart the definition of the spectral region to clip. Once the frame dimensions are optimized, the button *EXTRACT REGION* will clip the selected area. The name of the clipped region is required and the *SAVE* or the *SAVE and EXIT* buttons will save the new spectral portion in the XPS Sample.

The extracted regions can be utilized to estimate the concentration of elements not acquired in separated high-energy resolution spectra. Due to the different pass energy used, a specific procedure is applied to normalize for the different instrument efficiency.

3.6. CORE-LINES

Figures 3.4 and 3.5 show that XPS is sensitive to different chemical species providing distinct core-lines for individual elements allowing material speciation. However, the energy resolution used to acquire survey scans is low because of the broad acquisition energy range. In these spectra, any change due to bonding state of atoms cannot be appreciated. High-resolution core-line spectra must be acquired to reveal variations of the element binding energy, changes in the peak width and shape, modification of the valence states, and appearance of secondary structure as satellites. Formation of a chemical bond causes a deep rearrangement of the electronic charge around atoms leading to a change of their binding energy with respect to that of pure elements. This change is called *chemical shift* and its amplitude is linked to the oxidation degree of the atom. The chemical shift can be explained by the effective charge potential change on an atom [20–22]. Difference in the electronegativity of bonded atoms causes charge to flow from one to the other until equalization of the chemical potential [23]. The result is that an atom becomes positive, thus causing the BE to increase, while opposite occurs with the other atom. Figure 3.13 shows the change of the C 1s binding energy upon its oxidation state. In panel (a), carbon is bonded to an increasing number of fluorine atoms. As it can be observed, the C 1s binding energy rises linearly. In panel (b), carbon is bonded to different chemical elements possessing lower or higher Pauling electronegativity than that of C. The abscissa describes the electronegativity difference between carbon and that of the bonded element. The correspondent binding energy is reported on the ordinate axis. Also in this case, a linear relation is found. Data relative to Figure 3.13 are reported in Table 3.1.

A direct relation between the formation enthalpy and chemical shift is found for many oxides. A complete description of the theory behind the chemical shift including screening and relaxation effects can be found in [21]. The measure of the chemical shift needs the BE scale to be correctly aligned to a reference as described in Section 3.2. The interpretation of the chemical shifts is made using databases

---

TABLE 3.1

Trend of the C 1s Binding Energy as a Function of the C Oxidation State

Molecule	C1s BE (eV)	Molecule	χ_y (*)	$\chi_C - \chi_y$	C1s BE (eV)
($-CH_2-CH_2$)	285	MgC	1.31	−1.24	281.3
($-CHFCH_2-$)	286.78	ZrC	1.33	−1.22	281.6
($-CH_2-CF_2-$)	290.76	TiC	1.54	−1.01	281.6
$C_6H_5CF_3$	293.8	SiC	1.9	−0.65	282.7
CF_4	296.7	C=C	2.55	0	284.4
		CS_2	2.58	0.03	287
		CO_2	3.44	0.89	291.2
		CCl_4	3.16	0.61	292.4
		CF_4	3.98	1.43	296.7

Note: (*) χ_y represents the Pauling electronegativity of the atom bonded to carbon.

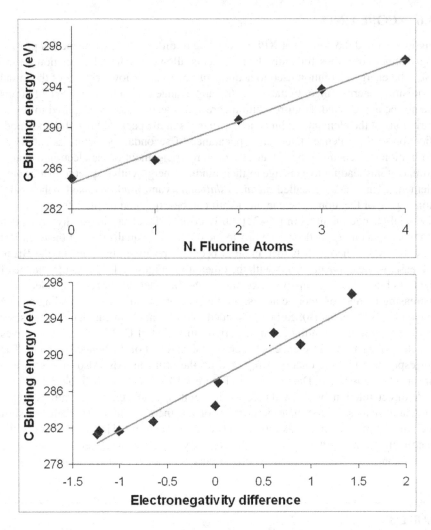

FIGURE 3.13   (a) Trend of the C 1s binding energy as a function of the number of fluorine atoms bonded. (b) Trend of the C 1s binding energy in binary compounds with increasing the Pauling Electronegativity difference between C and the bonded atom. Data are reported in Table 3.1.

and spectra acquired from certified samples [14, 24–28]. When the same element bonds to different chemical species, it forms bonds characterized by different charge exchange. Consequently, the core-line spectrum is the result of the assemblage of components at different chemical shifts as illustrated in Figure 3.14 for the carbon atom. Highly oriented pyrolytic graphite made by sp^2 carbon hybrids in Figure 3.14 displays a single peak at the lower BE corresponding to the carbon-carbon bonds in aromatic rings. This peak shifts to slightly higher BE in poly(ether ether ketone) (PEEK) and polyethylene-tereftalate (PET), because aromatic rings involve bonds

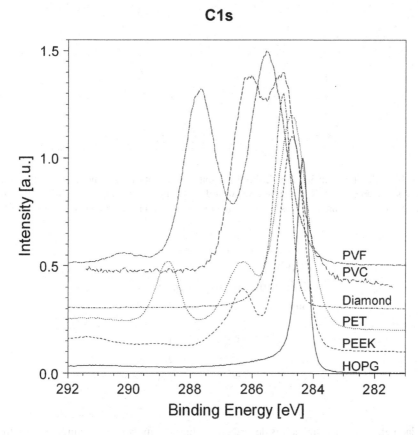

FIGURE 3.14 Different C 1s lineshapes resulting from different bonding: graphite, Polyetheretherketone, polyethylentereftalate, diamond, polyvinylchloride, and polyvinilfluoride.

with hydrogen in CH bonds. Spectra of these two polymers show also components at higher BE due to bonds with oxygen namely C–O–C and C=O in the case of PEEK and C–O–C and O-C-(C*=O) in that of PET [14]. Diamond is only made by sp^3 hybridized carbon atoms and displays a single peak corresponding to C-C single bond. Finally, the CH bond in spectra of polyvinylchloride (PVC) and polyvinylfluoride (PVF), is shifted at ~285.5 eV, because the presence of highly oxidizing chlorine and fluorine atoms. The second peak of their spectra corresponds respectively to C-Cl and C-F bonds [14]. Frequently the chemical shift is small and peak fitting is required to describe the energy and intensity of each "bond"-component.

3.6.1. SPIN-ORBIT SPLITTING

The C1s diamond and graphite spectra shown in Figure 3.14 are characterized by a single peak. In the case of emission from non-s-shells, the shape of the peaks changes and a doublet appears. Figure 2.15 shows an example of the 2p spectrum of silicon and of the 3d core-lines of niobium and silver.

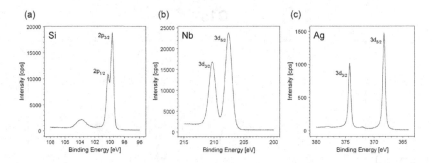

FIGURE 3.15   (a) Si 2p 3/2 and 1/2 the spin-orbit components are very near but still resolved by the spectrometer. (b) Nb 3d shows two well-distinct spin-orbit components 5/2 and 3/2. (c) The spin-orbit splitting increases by increasing the atomic number as shown for Ag 3d doublet.

TABLE 3.2

Spin-Orbit Splitting Parameters

Subshell	j values	Area ratio
s	1/2	
p	1/2, 3/2	1 : 2
d	3/2, 5/2	2 : 3
f	5/2, 7/2	3 : 4

Each electron in a subshell is described by the principal quantum number n, by the angular momentum l and by the spin s. The non-s-subshells are characterized by a non-zero value of l the angular momentum associated to the electron. The electronic state is described by the total momentum $j = |\ l \pm s\ |$ which for non-zero l assumes two different values. It comes out that the photoelectron spectra are split in two components corresponding to the two values of j which are described in Table 3.2. Figure 3.15 shows the effect of the L-S coupling and the generation of the different total angular momentum j in $2p_{1/2}$ and $2p_{3/2}$ of silicon (Figure 3.15A) and $3d_{3/2}$ and $3d_{5/2}$ of niobium and silver (Figure 3.15b, c). In the case of the 2p orbital, the splitting amounts to fraction of electronvolts. Figure 3.15b, c shows that the splitting increases to 2.72 eV for Nb and to 6.0 eV in the case of Ag 3d. In other words, the spin-orbit splitting increases with increasing the atomic number.

The dependence of the spin-orbit splitting ΔE from the atomic number z is explained considering that ΔE is proportional to the spin-orbit coupling ξ_{nl} which, on its turn, depends on the expectation value $< 1/r^3 >$ with r the average atomic radius. This results in an increasing energy separation ΔE with increasing the atomic number z for constant n while it decreases with increasing the value of the angular momentum l [29]. This is illustrated for the Au element in Figure 3.16.

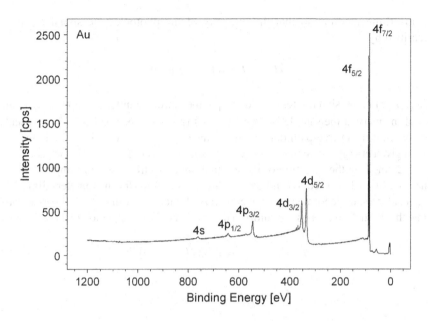

FIGURE 3.16 Wide spectrum of a pure gold sample. Increasing the value of the angular momentum *l* at constant *n* decreases the energy separation between the two spin-orbit spectral components.

3.6.2. Inelastic Mean-Free Path and Cross Section

3.6.2.1. Inelastic Mean-Free Path, Escape Depth, and Attenuation Length

There is a discrepancy between high penetration of X-rays in a solid and the depth sampled by an XPS instrument. This difference is related to the low probability that photoelectrons generated deeply into the material will reach the surface and possess kinetic energy high enough to leave the sample. It turns out that only photoelectrons produced just below the surface are able to escape the sample surface and this explains why XPS instruments are surface sensitive. This effect is due to the elastic and inelastic scattering processes affecting the transport of the electron to the sample surface. Increasing the depth increases the number of inelastic scattering processes leading to energy losses along the travel to the surface. Therefore, the probability that a photoelectron will be emitted in the vacuum is strictly linked to the depth where it is generated. Different terms are used to describe the effects of the inelastic scattering, namely the inelastic mean-free path, the attenuation length, and the escape depth.

Let us consider a number of photoelectron generated at a depth z below the surface. Their angular distribution will depend on the orbital involved in the photoemission process (Section 2.2.3.2 in [15]). If we assume straight trajectories, in a first approximation, electron attenuation is caused only by inelastic scattering. In this case, the attenuation is described as an exponential decay caused by an absorbing medium so that the intensity of the photoelectron signal I can be described by

$$I(z) = I_0 \exp[-z / \lambda \sin\theta] \qquad 3.1$$

where $I(z)$ is the signal intensity due to photoelectrons emitted at depth z, I_0 is the signal intensity at the sample surface $(z = 0)$ (see also Section 1.3.2). λ represents the inelastic mean free-path defined as the average distance traveled by an electron with a given energy between successive inelastic collisions. Finally, θ is the "take"-off angle defined by the direction of the emitted photoelectron and the sample surface. The relation (3.1) is used to estimate the thickness of thin (few nanometer) films. Let us consider a sample formed by an overlayer A of thickness d and a bulk B part and assume that both A and B possess uniform composition. Using equation (3.1), we obtain

$$I_A = I^0_A \left\{ 1 - \exp\left[-d / (\lambda_A \sin \theta) \right] \right\} \qquad 3.2$$

$$I_B = I^0_B \exp\left[-d / (\lambda_B \sin \theta) \right] \qquad 3.3$$

I_A and I_B are the measured intensities while I^0_A and I^0_B represent the intensities which were generated by uniform bulk A and B samples under identical conditions, λ_A and λ_B are the inelastic mean-free path (IMFP) of photoelectrons emitted by A and B core-levels. In general, λ_A is different from λ_B because the IMPF is function of the electron kinetic energy. In equation (3.2), we are describing the intensity I_A as a difference between that of a solid made of A and the contribution of the bulk below a film of thickness d.

A correct value of λ is essential not only to measure the thickness of an overlayer or to estimate the sampling depth in a homogeneous solid but also for a correct description of the surface composition of the material. For this reason, equations (3.2) and (3.3) were utilized to estimate the IMPF λ using samples made of films of known thicknesses on given substrates. However, these experiments led to IMPF estimates higher of about 30% than those expected from theoretical calculations because presence of elastic scattering increasing the electron mean-free path. Theoretical models show that different from the description given in equations (3.2) and (3.3), the signal attenuation does not follow an exponential decay. Last but not least, values of the attenuation length also depend on the experimental condition adopted.

These problems led to the definition of the *escape depth d* as the distance normal to the surface corresponding to a drop of a factor 1/e of the probability that a photoelectron leaves the surface without energy loss. As a consequence, 95% of the signal intensity comes from a depth equal to 3*d*. Introducing the escape depth in the exponential decay, for a take-off angle θ, the escape depth *d* may be described as $d = \lambda_{AL} \sin \theta$ where λ_{AL} defines the attenuation length. Using this new entity, $3d = 3\lambda_{AL} \sin \theta$ is assumed to be the sampling depth.

As seen in Section 1.3.2, experimental values of λ_{AL} were compiled in the past to retrieve the so-called *Universal Curves*, describing the behavior of the attenuation length as a function of the electron energies. It is intuitive that the attenuation length strictly depends of the characteristics of the material studied and in particular its density. For this reason, the *Universal Curves* were defined for organic and inorganic compounds [30].

The IMFPs for elements with energies between 1 eV and 10,000 eV above the Fermi level are described by

$$\lambda_{AL}\,(\text{monolayers}) = \left[538/E^2 + 0.41\,(a\,E)^{0.5} \right] \qquad 3.4$$

where a is the mean atomic distance ("monolayer thickness") in nanometers which can be calculated from

$$a = \left(M/\rho N_{Av} \right)^{1/3} \qquad 3.5$$

where a is measured in meters, M is the atomic mass in kg, ρ is the material density in kg/m³, and N_{Av} is the Avogadro number corresponding to $6{:}02 \bullet 10^{26}$ kmol⁻¹.

For inorganic solids, λ_{AL} is described in monolayers by

$$\lambda_{AL}\,(\text{monolayers}) = \left[2170/E^2 + 0.72\,(a\,E^{0.5}) \right] \qquad 3.6$$

Finally for organic compounds, the experimental results were fitted using the function:

$$\lambda_{AL}\,(\text{nm}) = (10^3/\rho)\,(49/E^2 + 0.11\,E^{0.5}) \qquad 3.7$$

More recent papers have been published on the effective attenuation lengths [31–34]. A refined description of the IMPF is the TPP-2M model due to Tanuma, Powell, and Penn, who described the attenuation of the photoelectron intensity using a modified Bethe equation to describe the inelastic scattering [35]. For electron energies between 10 eV and 2,000 eV for any material following the TPP-2M equation, the IMPF is described by

$$\lambda_{IMPF} = E/\left\{ E_p^{\,2}\left[\beta\,\ln(\gamma E) - (C/D) + (D/E^2) \right] \right\} \qquad 3.8$$

where E is the electron energy in eV, and $E_p = 28.8\,U^{0.5}$ with $U = (N_v\rho/M)$ is the free electron plasmon energy in eV. N_v is the number of valence electrons per atom, M is the atomic or molecular mass, and ρ is the density in g/cm³. Parameters β, γ, C, and D are obtained empirically considering other material properties:

$$\beta = 0.10 + 0.944/\left[E_p^{\,2} + E_g^{\,2} \right]^{0.5} + 0.069\rho^{0.1} \qquad 3.9$$

$$\gamma = 0.191 \, \rho^{-0.5} \qquad\qquad 3.10$$

$$C = 1.97 - 0.91U \qquad\qquad 3.11$$

$$D = 53.4 - 20.8U \qquad\qquad 3.12$$

where E_g is the band-gap energy (in electron volts). Equation (3.8) was used to estimate the IMPF in elements and inorganic [36] and organic compounds [35, 37]. A database providing values of IMPF useful for surface analysis performed either by XPS or by AES is also available at the NIST web page [38]. Values of IMPF were estimated using three different kinds of sources: (i) calculated IMFPs from experimental optical data; (ii) IMFPs measured by elastic-peak electron spectroscopy for some elemental solids (only for a limited number of materials); (iii) for all materials, the IMFPs were obtained from predictive formulae. Finally, more extended description of the electron transport in solids and the electron attenuation length may be found in [39] and [40].

3.6.2.2. Cross Sections and Sensitivity Factors

For a homogeneous solid composed by atoms A with density ρ_A illuminated by X-photons of energy hv and intensity I_{hv}, it is possible to describe that the photoelectron current $I_{A,i}$ generated ionization of a core-level i $\left(i = 1s, \, 2p, \, 3d\ldots\right)$ considering the integral of the spatial distribution of excitation and emission [41, 42]:

$$I_{A,i} = \Delta\Omega \, / \, 4\pi \int_{0}^{\infty} I_{hv}\left(\alpha, \, z\right)\sigma_{A,i}W_{A,i}\left(\beta_{A,i}\,Y\right)\rho_A\left(z\right)\exp[-z/(\lambda_{A,E}\sin\theta)] \, dz \qquad 3.13$$

$\Delta\Omega$ is the acceptance solid angle of the XPS analyzer, α represents X-ray incidence angle, $\sigma_{A,I}$ is the ionization cross section of orbital i of element A, and $W_{A,i}\left(\beta_{A,i}\,\Psi\right)$ is the angular asymmetry factor [43]. The remaining term describes the attenuation of the signal due to energy losses as described by equation (3.1) with $\lambda_{A,E}$ being the attenuation length of photoelectrons from core-level i of element A. Equation (3.13) is based on the assumption that any diffraction effect caused by crystalline structure is absent.

Let us consider the cross section $\sigma_{A,i}$ which was calculated by Scofield [44] treating electrons relativistically in a Hartree-Slater central potential for all elements. The calculation of the cross section is made assuming the electrons be immersed in the same central potential both before and after the absorption of the photon. All contributing multipoles and retardation effects are included in the treatment of the radiation field. The cross sections calculated by Scofield do not consider screening effects leading to intrinsic plasmon losses. More precise evaluation of the element photoemission efficiency was done by Wagner [45], who obtained the relative elemental sensitivity factors (RSFs) by experimental measurements. RSFs were obtained by analyzing 135 compounds of 62 elements. RSFs are obtained as intensity ratios of spectral lines with F 1s as a primary standard and with K $2p_{3/2}$ as a secondary standard. The experimental RSFs against binding energy on a log–log graph were fitted using curves obtained from equivalent plots of theoretical cross sections. The curves provide the RSFs for both the more intense and the secondary transitions of all elements.

Recently, a more precise evaluation of RSFs [46] was done using both Al and Mg Kα radiation considering that (i) Wagner's work does not provide information if the bulk composition of the standards is uniform in the XPS analysis depth ensuring unreacted surfaces; (ii) no specific information is provided about the background subtraction rendering difficult the reproducibility of data. Results are shown in Figure 3.17. Direct use of the RSF listed in [45] or [46] is possible only if the user possesses the same XPS instrument. In other cases, the different analyzer transmission function and detector efficiency at various BE may introduce consistent errors. Generally the instrument manufacturers provide their own RFSs calibrated for the analyzer. In *RxpsG*, the value of the RSF is tabulated for the Scienta and Kratos instruments. When a spectrum is loaded in the *RxpsG* memory, the correct RSF are automatically associated to all the core-lines. In the case of more than one RSF, all the values are present in the *RxpsG* element Tables, and the user is asked to associate manually one of these tabulated values to the actual core-line.

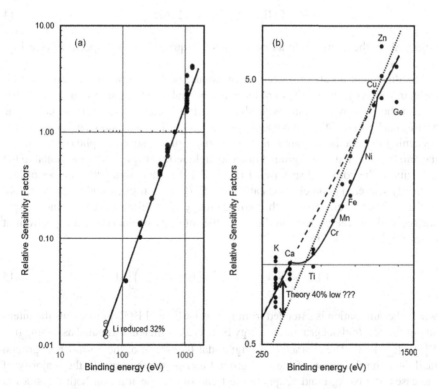

FIGURE 3.17 Trend of relative sensitivity factors obtained experimentally for all elements using standard compounds with known composition.

Source: Reprinted with permission from [46]

3.6.3. BACKGROUNDS

Generally three different approaches are utilized for background subtraction: *Linear*, *Shirley*, and *Tougaard*. The background is generated by the sum of scattered electron exiting the material surface at the different energies. Then background depends on the material composition and on the element distribution with depth. The description of the physics behind the generation of the background is not easy and still partial. None of the three backgrounds offer a true solution for any kind of material, but for more or less restricted classes of materials, their application gives reasonable results.

When the background end points are unleveled, linear background is inappropriate, since it may cause large peak-area changes. *Linear* backgrounds are more suitable for peaks placed on a flat background as displayed in Figure 3.18a. Generally this occurs in non-conducting materials characterized by a large band gap. In this case, a change of the background end-point positions does not influence the value of the peak area. If $S(E_i)$ represents the measured spectrum and $F(E)$ indicates the background corrected spectrum, in the energy range $E_{min} < i < E_{max}$ for a linear background

$$F(E_i) = S(E_i) - k\Sigma_i \Delta E \qquad 3.14$$

where ΔE is the energy step adopted for the acquisition of the spectrum (see Figure 3.18a).

Metallic materials are characterized by intense peaks generating a cascade of photoelectron energy losses. This produces the typical asymmetric tail on the high BE side of the peak with pronounced sharply rising backgrounds. In these cases, the *Shirley* background [47] is more appropriate than the linear as shown in Figure 3.18b. Assuming that the background stems from inelastically scattered photoelectrons, in the *Shirley* model, the background intensity at kinetic energy E_i is proportional to the intensity of the corrected spectrum $F(E < E_i)$. In other words, the number of inelastically scattered photoelectrons at energy E_i (Shirley background) is proportional to the number of electrons with binding energy $E < E_i$ emitted from the sample surface without energy loss. In this case, the intensity of the corrected spectrum at energy E_i is described by

$$F(E_i) = S(E_i) - k_i \Sigma_j \left[S(E_j) - B(E_j) \right] \Delta E \qquad 3.15$$

where the summation is extended from $E_{min} < j < E_i$, and $B(E_j)$ represents the intensity of the *Shirley* background at energy E_j. k_i is a coefficient estimated assuming that $F(E = E_{min}) = 0$. The characteristic sigmoidal-like shape of the *Shirley* background facilitates the definition of the background end-points, because in the majority of the cases, the background adapts to the trend of the spectrum on both the sides at high- and low-binding energy edges. It may happen that the pure Shirley background crosses the spectrum. In this case, the *2P.Shirley* may solve the problem. It can be demonstrated that the *Shirley* background may be described with a *Shirley cross-section* function K_S [48]

$$K_S(T) = B_S * T / \left[C_S + T^2 \right]$$ 3.16

where $T = E' - E$ describes the energy loss. K_S describes the probability that an electron of energy E propagating in the material shall lose energy $E' - E$ per unit path length. B_S coefficient is the "scattering intensity" and is automatically computed to make the background to assume the spectral intensity at high BE (low KE) edge of the region where the background has to be defined. C_S describes the "distribution width". $C_S = 2500$ was evaluated for a class of elements with core-lines at different BE.

The general description of the *Shirley* background is:

$$S(E) = \int_E^\infty K_S(E'-E) \left(J(E') - S_0(E') \right) dE'$$ 3.17

where $J(E)$ is the measured spectrum at energy E and So represents the contribution of photoelectrons generated at higher kinetic energies [48]. Application of eq. 3.17

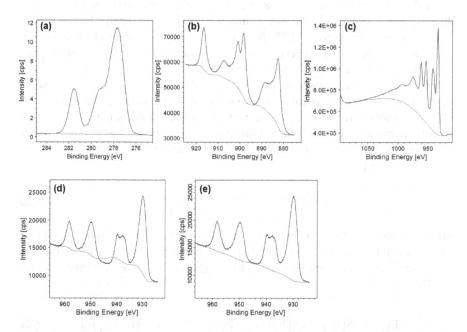

FIGURE 3.18 (a) Linear background applied to a C 1s spectrum showing leveled endpoints. (b) Shirley background applied to Ce 3d spectrum showing a marked different intensity at the endpoints. (c) The Cu 2p core-line with its loss features. A Tougaard background over an extended energy range is used in this case. (d) Shirley background crosses the CuO spectrum. (e) 3P Shirley baseline correctly describes the background.

simplifies the computation of the *Shirley* background. However, this method can cause the divergence of S(E) when the background crosses the spectrum. This sometimes can occur as in the example of Figure 3.18d. In this case, the *3P.Shirley* can be used. This algorithm uses a modified expression of the *Shirley cross-section* function and can be used when also the *2P. Shirley* is inappropriate. Now the Shirley cross section is multiplied by the factor $1 - \exp\left[-D_s\left(E' - E\right)\right]$ which concurs to lower the background intensity below the spectral features. The modified *Shirley* background now is described by:

$$S(E) = \int_E^\infty \left[B s\ T\ \left(1 - e^{-Ds\ T}\right)/\left(Cs\ +\ T^2\right)\right] * J(E)dT \qquad 3.18$$

where $T = E' - E$ is the energy loss. The result is shown in Figure 3.18e. Finally in *RxpsG* is also implemented the *LP.Shirley* which combines the *Shirley cross-section* function with a linear weakening polynomial introduced by Bishop (see ref. [49])

$$K_S(T) = B_S T\left(1 - mT\right)/\left(C_S + T^2\right) \qquad 3.19$$

where $T = E' - E$ is the energy loss and m is a parameter defined by the user. The Bishop polynomial weakens the background contribution as the energy loss increases.

The Tougaard background [50] is computed using an universal cross-section function [51] describing the energy-loss processes due to the interaction of the photoelectron with the electric field developed by the material atoms charges whose distribution adapts in response to the presence of the moving electron. The inelastic scattering cross section was calculated on the basis of the dielectric function which may vary in solids. However, it was found that, for most of them, the product of the inelastic electron mean-free path and the inelastic scattering cross section is rather weak and can be described by a Universal Cross Section, a function of the energy loss $T = E' - E$ [51]

$$K_T(T)\ =\ BT/\ \left[C + T^2\right]^2 \qquad 3.20$$

where K_T describes the probability that an electron of energy E through scattering processes will lose energy $E' - E$ per unit path length traveled in the solid. According to Tougaard, the background generated by energy losses may be described as [15, 51, 52]

$$F(E_i) = S(E_i) - \Sigma_j\, S(E_j)\Delta E\ \bullet\ B(E_j - E_i)/\left[C + (E_j - E_i)^2\right]^2 \qquad 3.21$$

where the summation is extended from $E_{min} < j < E_{i-1}$, $S(E_i)$ is the intensity of the measured spectrum at energy E_i, ΔE is the energy step used during acquisition, and B and C are fitting parameters (for this reason, this background in *RxpsG* is named

2.P Tougaard). However, it was found that B = 681.2 eV² and C = 355 eV² give reasonable results for 59 elements [41, 52]. Equation (3.20) describes the intensity of the background at a given energy as a weighted integral of the spectral intensities at lower BEs (higher kinetic energies). An example of Tougaard background is shown in Figure 3.18c for an extended Cu 2p core-line including the loss structures at high BE.

3P.Tougard is defined for non-metallic elements; it was proposed a modified universal cross section K_{TM} [53, 54] defined by

$$K_{TM}(T) = B_T T / \left[\left(C_T + T^2 \right)^2 + D\ T^2 \right]$$ 3.22

where C_T and D_T are constants characteristic of the solid. The parameters were evaluated by Tougaard for a class of elements and are fixed to $C_T = 551$ eV and $D_T = 436$ eV. Again B_T is computed automatically in order to make the background intensity equal to that of the core-line at the high BE (low KE) edge. Then the background-subtracted spectrum F(E) at energy E may be expressed as:

$$F(E) = J(E) - B_T \int_{E}^{E_{max}} K_{TM}(E' - E)\ J(E')\ dE'$$ 3.23

where J(E) is the measured spectrum.

4P.Tougard is intended for general use and is derived adding a further modification to the previous version of the universal cross section K_{TM} in equation (3.23) introducing the parameter C' so that now the cross section becomes a four-parameter cross section K_{T4}:

$$K_{T4}(T) = B_T T / \left[\left(C_T + C'T^2 \right)^2 + D\ T^2 \right]$$ 3.24

where C_T and D_T are fixed constants assuming same values $C_T = 551$ eV and $D_T = 436$ eV as in the 3P.Tougaard K_{TM} cross-section function. Then the background-subtracted spectrum F(E) at energy E may be expressed as:

$$F(E) = J(E) - B_T \int_{E}^{E_{max}} K_{T4}(E' - E)\ J(E')\ dE'$$ 3.25

where $J(E)$ is the measured spectrum at energy E. The parameter C' is manually adjusted to obtain the best background below the spectral data.

The principal difference between Shirley and Tougaard backgrounds is the higher level of control offered by the second approach. As in the previous cases, the definition of the endpoint at low BE (high kinetic energy) generally is easy, because the background in that region is flat and the spectrum sharply increases at the onset of the peak. More difficult is the definition of the end point at high BE because of the presence of loss features as plasmons, shake-up features, or in presence of spectral components on the high BE sides of the peak. The Tougaard background definition is more accurate than the previous background models. However, the definition of the background using the Tougaard Universal cross-section function requires a large energy range extending some 50 eV to the higher binding energy side of the core-line analyzed to get the data needed for a reasonable fit. Generally the energy range of the acquired spectra is much narrower making the Tougaard background less frequently used with respect to the Shirley approach which generally is sufficiently accurate [39, 55].

3.6.4. CORE-LEVEL LINESHAPES

In photoemission processes, the natural width of the spectrum is originated by the uncertainty in the lifetime of the core-hole. The core-hole corresponds to an excited state, and because this excited state decays exponentially in time, this generates a line with Lorentzian shape as a function of the frequency [56]. The natural line broadening ΔE_0 originates from the lifetime of the core-hole generated by photoemission and can be estimated by the Heisenberg principle as

$$\Delta E_0 = h/\tau = 4.1 \times 10^{-15}/\tau \qquad\qquad 3.26$$

where h is the Planck constant expressed in eV s, while τ represents the lifetime of the core-hole in seconds. A typical lifetime is about 10^{-14} s corresponding to a line width of 0.4 eV, a value measured for the Ag $3d_{5/2}$ peak. Commonly, the core-line is affected by other sources of broadening. Among these, there are various kinds of final-state effects involving excitation of many electrons [57, 58], vibrational excitation [59] which are added to the instrumental broadening deriving from the excitation line-width and from the analyzer. There is not a universal lineshape which can describe any kind of core-line peak. Most of the analyses are carried out trying a simple Gaussian:

$$I_G(E) = 1/\sigma(2\pi)^{0.5}\exp[-0.5(E - E_0)^2/\sigma^2] \qquad\qquad 3.27$$

with σ = full with at half maximum of the peak, E_0 is the peak position, or a Lorentzian function

$$I_L(E) = 1/\left\{1+4*\left[(E-E_0)/\gamma\right]^2\right\} \qquad\qquad 3.28$$

with γ representing the FWHM of the peak, or using a convolution of these two function as the Voigt lineshape

$$I_V(E,\sigma,\gamma) = \int_{-\infty}^{\infty} G(E',\sigma)\, L(E-E',\gamma)\, dE' \qquad\qquad 3.29$$

where G and L represent the Gaussian and Lorentzian lineshapes, and looking at the best fitting results. The Voigt function is somewhat more realistic accounting for both the natural (Lorentzian) and instrumental (Gaussian) contributions to the spectral broadening [60, 61]. All these lineshapes result to be ineffective when core-lines deriving from conducting materials are analyzed. In these cases, the XPS peak is asymmetrically skewed by a tail at higher binding energies. This distortion is caused by many-body interactions of the photoelectrons with the free electrons at the Fermi Edge typical of the conducting materials. The theory behind this effect was described by Doniach and Sunjic (DS) [62] leading to a lineshape function which is indicated using their names

$$I_{DS} = \Gamma(1-\alpha) \cdot \cos[\pi\alpha + \theta(\varepsilon)] / [(\varepsilon^2 + \gamma^2)^{(1-\alpha)/2}] \qquad 3.30$$

with

$$\theta(\varepsilon) = (1-\alpha) \arctan(\varepsilon / \gamma) \qquad 3.31$$

where ε is the energy measured from the core-level position, γ is the FWHM of the natural (Lorentzian) line, and α is the characteristic asymmetry factor $0 < \alpha < 0.5$ [63] which is determined by the phase shift of the l orbital for scattering of conduction electrons from the hole potential [62]

$$\alpha = \Sigma_j (2l+1)(\delta / \pi)^2 \qquad 3.32$$

Here the summation is extended to all values of angular momentum starting from $l = 0$. α results to be larger for s-type screening (e.g., 0.20 for Na, 0.13 for Mg, and 0.12 for Al).

The DS function reflects the probability of exciting free electrons, as a function of the energy E. This probability is proportional to the density of stated $D(E_F)$ around the Fermi level. This occurs in conducting samples since conductivity is associated to overlap of the valence and conduction bands with absence of a band gap. In addition, as the interaction between the exiting photoelectron and the Fermi electrons increases, the core-lineshape changes from pure Lorentzian (no interaction) to a more and more asymmetric DS function as represented in Figure 3.19.

Appropriate lineshapes for fitting experimental peaks are summarized in Table 3.3.

TABLE 3.3
Suggested Lineshapes for Given Material Species

	Insulators, semiconductors, and metals with low $D(E_F)$	Metals with high $D(E_F)$
Basic Lineshapes	Lorentzian	Doniach-Sunjic
Lineshapes in presence of broadening	Voigt, Gaussian	Doniach-Sunjic + Gaussian

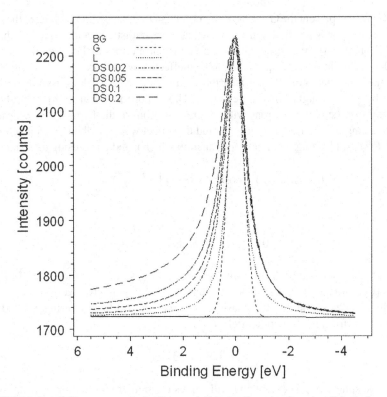

FIGURE 3.19   BG represents the background, G is the Gaussian function, L is the Lorentzian function, and DS is the Doniach Sunjic lineshapes using the asymmetry factor α indicated in the legend. The lineshapes are plotted as a function of the energy loss.

3.7.   CORE-LINE SECONDARY STRUCTURE

Photoemission process causes a perturbation of the electronic structure of the atom emitting the electron. Final state effects refer to any kind of intra- or inter-atomic interaction deriving from the formation of a core-hole. These interactions can be classified in two groups: (a) polarization induced by core-hole formation; (b) core-hole-induced excitation of outer electrons. In the first case the creation of a core-hole generates an instantaneous electric field which may influence the spectral response of the emitting atom. In the second case, the rearrangement of the electronic charge of the atom or of the environment is due to the coupling between the photoelectron resulting in an energy transfer which excites charges in a higher energy level. This occurs within the timescale of the photoemission process, thus resulting in the modification of the spectral features.

These final state effects lead to the generation of core polarization effects, photoelectron peak asymmetries, and Satellites and Plasmon loss structures, and photoelectron-induced Auger electron features.

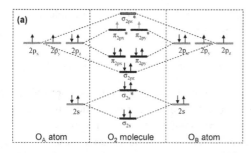 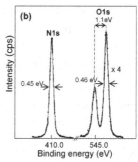

FIGURE 3.20 (a) Scheme of the energy level in single O atoms and how they transform in the composition of the O_2 molecules. The upper levels π_{2px}^* and π_{2py}^* show unpaired electrons leading to a value of the total spin S = 1. The energy levels of the N_2 molecule are similar to that of O_2 apart from the absence of the unpaired electrons in the upper excited level. The total spin of N_2 is S = 0. (b) The different value of the total spin results in a single peak for N 1s (S=0) and a doublet for O 1s (S = 1).

Source: Reprinted with permission from [64]

3.7.1. Core-Hole Polarization

Photoemission from an atom core-level generates always a final state with spin and angular momentum. If the analyzed system possesses an open valence-state config-uration corresponding to unpaired electrons, the corresponding state is characterized by an orbital momentum and a non-zero spin. This spin can couple with the angular momentum and spin of the core-hole leading to a number of different final states. A good example is the oxygen O_2 molecule which is characterized by a value of the total spin S = 1 as illustrated in Figure 3.20.

In Figure 3.20, the photohole generated in the O_2 molecule can couple to the valence total spin S bringing to a splitting of the core-line in two components corre-sponding to the value J = 1/2 and J = 3/2 separated by 1.1 eV. Different is the case of the N_2 molecule. The N atom possesses one electron less than the O atom. As a consequence, the electronic structure of this molecule is similar to that of the O_2 mol-ecule without the electrons in the higher electronic level (indicated in with different color Figure 3.20) and with the π_{2p} levels placed below the σ_{2p}. In this case, the N_2 molecule has a total spin = 0 and a single peak is then generated (see Figure 3.20). Core polarization occurs also in solids, for example, in 3d transition elements [65, 66] or rare earth elements [67, 68]. Exact calculation is rather complex due to addi-tional effects. Among other, the creation of the core-hole will generate a coulombic potential attracting the remaining electrons toward the nucleus. This increases the nucleus shielding, and the binding energy of the exiting photoelectron is reduced. But the extent of the reduction will depend on the system electronic structure and polarization which, in their turn, are depending on the initial kind of bonding. The estimation of the core-hole energy shift was done for the Cl 2s and Na 1s [69, 70] and amounts to ~1.5 and ~2.5 eV, respectively. Presence of the core-hole polarization induces multiplet splitting via exchange interaction.

3.7.2. MULTIPLET SPLITTING

If an atom displays a net spin S originating from the generation of a core-hole or from the unpaired electrons in the valence band, an exchange interaction may take place. The multiplet splitting is created by an exchange interaction which resembles to some extent the spin-orbit interaction. Let us consider first the easier case of *s* core-levels with null angular momentum. In this case, the exchange interaction occurs between the unpaired electron spin left in the core-level and the spin of unpaired valence electrons of the atom. If the spins are parallel, the exchange interaction between spins leads to a lower binding energy with respect to that obtained for antiparallel spins. This results in a splitting of the core-level in two components and the energy separation between the two peaks is described by the exchange interaction given by

$$E = (2S + 1) \ K_{s,v} \qquad\qquad 3.33$$

where S represents the total spin generated by the unpaired valence electrons, and $K_{s,v}$ represents the exchange interaction between the initial core-level $|s\rangle$ and the valence electrons $|v\rangle$.

Let us consider a non-s core-level as initial state; in this case, the description of the exchange interaction is more complex because it involves also the non-zero angular momentum. The energy separation among the multiplet of features generated may be described by [13].

$$E_{c,v} = \left[\left(2(S_v + s_c) + 1 \right) / (2l+1) \right] K_{c,v} \qquad\qquad 3.34$$

where $K_{c,v}$ is the multiplet splitting exchange integral, and S_v and s_c indicate the total spin of the valence and core level, respectively. The inclusion of the core-level spin s_c was suggested [71] to better describe the multiplet $E_{c,v}$ previously developed to describe the effect of a valence hole, when core-level are involved. The exchange integral $K_{c,v}$ increases if both the core and the valence levels are identified by the same principal quantum number n while it decreases as the interacting core electron BE increases (value of the associated angular momentum decreases leading to deeper atomic levels). As a consequence of this, it is easily understandable that a relation of $K_{c,v}$ with the BE of the core-level affected by the multiplet splitting. As an example, this relation was found for the first row transition metal ions where the value of the energy separation $E_{c,v}$ of the *3s* core level can be described as a function of the BE_{3s} by [72]

$$E_{c,v} = \left[2(S_v + s_c) + 1 \right] / (0.6 \ BE_{3s})^{1/9} \qquad\qquad 3.35$$

A clear description of calculation of the coupling of the core-hole and unpaired valence electrons is described in [73] and shown in Figure 3.21.

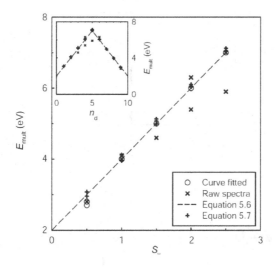

FIGURE 3.21 Multiplet splitting energy $E_{c,v}$ versus S_v. In the inset is shown $E_{c,v}$ versus the number of electrons in the transition metal ion 3d level (n_d). The dashed lines represent equation (3.34) with the sum of s_c and S_v represented by S. Also shown is the relation between $E_{c,v}$ and the BE_{3s} as relayed through equation (3.35).

Source: Reproduced with permission from [72]

3.7.3. SHAKE-UP SATELLITES

The creation of a core-hole by photoemission is accompanied by a decrease of the electron charge shielding and to a correspondent increase of the nuclear coulombic field. This causes a strong perturbation leading to a reorganization of the valence electron distribution. In order to minimize the total energy of the system, the remaining valence electrons "relax" and the associated relaxation energy is of the order of ~ 10–20 eV. This reorganization of the valence charge has an effect also on the BE of the photoemitted electron. To better explain this effect, let us consider the sudden approximation (Koopmans' theorem) where any relaxation effect next to the photoemission is disregarded. In this hypothesis, the energy of the photoelectron is simply given by the excitation energy E_k where k indicates the orbital from which the electron is excited. In this case the spectrum is formed by a single Lorentzian peak falling at energy E_k. However, immediately one realizes that the system will respond to the photoionization process with a reorganization of the electron charge left on the atom. Because these effects occur on a timescale comparable to that relative to the photoemission, the sudden approximation is no more valid, and the generation of a k core-hole will lead to a series of effects on both the photoelectron energy and the generation of new spectral features. The different behavior of the system in the case of sudden approximation $V = 0$, where V represents the potential generated by the core-hole or, in the more realistic case of $V \neq 0$, is represented in Figure 3.22

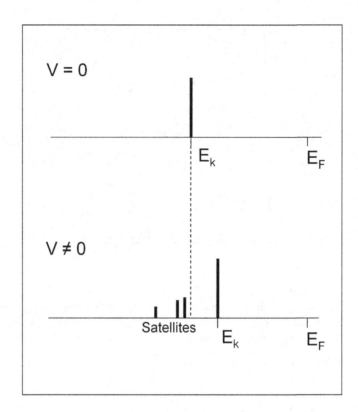

FIGURE 3.22 $V = 0$ sudden approximation (Koopmans) coulomb and exchange interactions are disregarded. The photoelectron energy E_k corresponds to the energy of the k level where the hole is created. $V \neq 0$ represents the presence of interactions. The photoemission generates secondary structures, the satellites, and a back-shift of the photoelectron BE. The center of gravity of the two spectra remains unchanged.

In both the cases, the "baricenter" of the spectrum remains unchanged [69].

Apart from the BE shift of the primary spectral structure, in presence of coulomb and exchange interactions, the creation of a photohole ends up also with another effect related to the non-zero probability of exciting the remaining electron charges as shown in Figure 3.22 for $V \neq 0$. The excitations are made at expenses of the exiting photoelectrons and appear in the spectrum as structures at higher BE. The spectrum is then formed by a "principal" line associated to the photoemission processes without energy loss and a number of secondary structures, the satellites, which describe photoelectrons at lower energy due to excitations of the valence charges. These consist in transitions of the valence electron from a bonding orbital to an excited level. Two kinds of satellites are formed: the shake-up structure corresponds to the excitation of a valence electron into an antibonding orbital simultaneously to the formation of the core-hole. This is the case of aromatic organic compounds where the shake-up transitions are associated to $\pi \rightarrow \pi^*$ excitations representing the HOMO and LUMO orbitals involved as illustrated in Figure 3.23a. Another typical example regards the

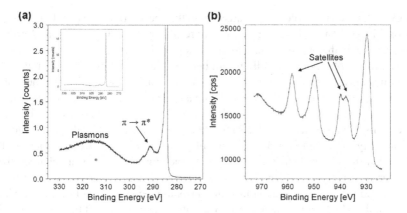

FIGURE 3.23 (a) C 1s spectrum derived from highly oriented pyrolytic graphite, loss features are indicated. In the inset, the complete spectrum is shown. (b) Cu 2p core-line acquired in CuO showing the presence of intense satellites.

transition metals and rare earth compounds possessing unpaired electrons in their 3d and 4f orbitals, respectively. Thanks to the open shell configuration, the final state of these materials displays a significant increase of charge [49]. This explanation also helps understanding why Cu 2p spectrum with a $3d^{10}$ closed shell electronic configuration (remember that the shell structure of Cu is [Ar] $3s^2$ $3p^6$ $3d^{10}$ $4s^1$, the latter electron lacking after photoemission and relaxation) does not show any satellite structure. Differently this latter appears, for example, in the Cr_2O_3, Fe_2O_3, and CuO spectra where excitations are possible. In the case of CuO, the shake-up satellites shown in Figure 3.23b derive from the excitation of an O_{2p} valence band electron of CuO into the unfilled 4sp states in the conduction band near the Fermi Edge E_F [74].

3.7.4. Shake-Off Satellites

Shake-off processes are similar to the shake-up, but in this case, the energy released to the outer electrons is high enough to excite them from valence levels above to vacuum level into unbound continuum states. The shake-off satellites are rarely seen as discrete peaks. More frequently, they result in a contribution to the core-line width or to the core-line background on the higher BE side of the main photoelectron peak.

3.7.5. Plasmon Loss Features

Solids exhibiting a certain density of free electrons around the Fermi level frequently display characteristic loss features related to plasmon formation. Plasmons are collective oscillations of the quasi-free electrons in the proximity of the E_F level. These oscillations occur at a frequency ω and at its harmonics of higher order, which are characteristic of the material. A contradistinctive characteristic of the plasmons is the sequence of features at BEs increasing in steps of $\hbar\omega$. Photoelectrons possessing energy equal to $n \cdot \hbar\omega$, where $n = 1, 2, 3 \ldots$, can excite plasmons and experience an energy loss. Good examples of this are seen in the spectra of Mg, Al, and Si. Plasmon loss oscillations on the high BE side of Si are illustrated in Figure 3.24a. With oxidation generally the plasmon oscillations disappear. However, there are cases, as that of TiO_2, where plasmons are still present because the formation of a core-hole. This can induce excitation of valence electrons into the conduction band. Then they move in response to the sudden variation of the charge density occurring on photoelectron production leading to excitation of collective oscillations in the conduction band.

At the material surface, the electronic structure is influenced by the solid–vacuum discontinuity. As a consequence, diverse plasmon oscillations are generated in the bulk and at the material surface. Bulk and surface plasmon losses are characterized by different principal frequencies and then fall at different energies with different energy periodicity. Figure 3.24b illustrates an example in the case of Mg. On the high BE tail of the Mg 2p core-line, intense loss features occur. Each of them is formed by two components assigned to surface S and bulk B plasmons.

Plasmons resulting from direct excitation processes by the photoelectron are referred to as *intrinsic plasmons*. In other cases, the excitation may occur as a secondary effect of the formation of a core-hole and the sudden change in charge density or may be generated by the passage of fast charged particles (electrons resulting from photoelectron emission). In these cases, plasmons are referred to as *extrinsic plasmons*.

Whereas the extrinsic plasmons contribute to the background, intrinsic plasmons result in a lowering of the emission intensity, thus requiring a correction of the Scofield ionization cross section. On the other hand, they are not included in the calculation of the inelastic mean-free path. Since with formation of oxides the plasmons disappear as already noted, different cross-section values should be utilized for pure elements or their oxides. To avoid the use of different cross-section values, plasmon-loss features must be added to the main core-line peak in the quantification process. Similar procedure holds for estimating the attenuation length values through the layer thickness method.

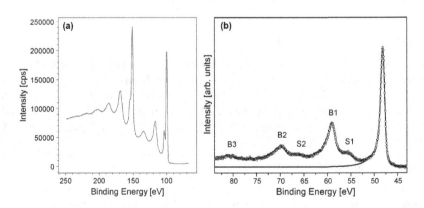

FIGURE 3.24   (a) Si 2p and 2s (at higher BE) displaying a series of ordered oscillations corresponding to the excitation of the plasmons. (b) Mg 2p core-line followed by surface (S) and bulk plasmon components indicated with S1, S2 and B1, B2, B3, respectively (S3 too weak to be observed).

Source: Reprinted with permission from [75]

3.8.   CORE-LINE SPECTRAL ANALYSIS

This section is dedicated to the illustration of the *RxpsG* functions to perform the spectral analysis. The logical sequence of the operation is:

Baseline and *Region to fit* definition → Selection/location of the fit components → Constraints on fitting parameters → Best fit → Bond assignment

Let start with the creation of the baseline and the definition of the fitting components.

3.8.1.   ANALYZING CORE-LINES

The *Analysis* option of the main menu (Figure 3.25) contains all the functions to perform the spectral analysis of a core-line. The *Analyze* option is accessible from the *Analysis* menu containing all the options useful for spectral analysis and data manipulation and reduction.

Selection of the *Analyze* option will open the relative interface to select the various operations needed for the spectral analysis. The GUI is illustrated in Figure 3.26.

At the top left of the GUI, you can select the *XPS-Sample*. All the *Analyze* options are enabled by selecting the core-line to analyze on the top right side. In the graphic window, the selected core-line will appear. At the ends of the spectrum, there are two active markers (cross circles). The markers must be moved in the positions of the edges of the so-called *Region-to-fit*. This region defines the portion of the spectrum which will be analyzed. Upon selection, the background will be placed under the spectrum. The user can readily modify the background edges or change the background lineshape by selecting the desired function. The selection of the appropriate lineshape depends on the nature of the material analyzed. Generally, linear background can be selected if the spectral intensity at the ends of the *Region-to-fit* is low or in the

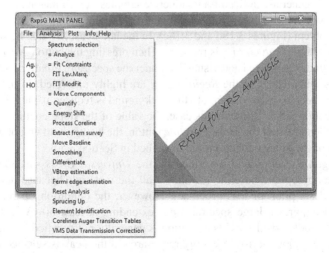

FIGURE 3.25   The *Analysis* option of the main menu contains a number of options useful for spectral analysis, manipulation of core-lines, smoothing, extract data from wide spectra, estimation of the position of the Fermi level, and chemical element identification.

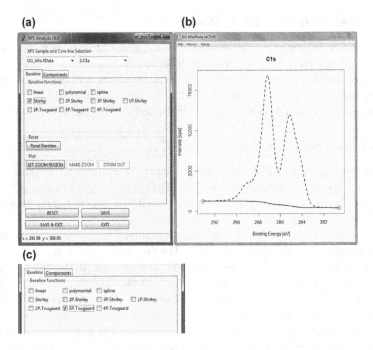

FIGURE 3.26   (a) The *Analyze* interface together with the active graphical window (b) and the list of available baseline functions (c).

cases where a change of the ends position does not introduce appreciable variations of the area contained between the spectrum and the defined background. The *Shirley* background is generally utilized for conductive samples. It may happen that the background crosses the spectrum. In these cases, *2P.Shirley* or *3P.Shirley* can be used to solve the problem (Figures 3.18d and 3.18e). In the latter case, a value between 0 and 1 of the distortion parameter *Ds* is required. Then pressing the button *Make Baseline*, the background is generated and visualized under the spectrum. Finally, the *LP.Shirley* can be used when edges of the *Region-to-Fit* are highly unleveled. In this case, the *Shirley cross section* used to compute the background is reduced using a linear function (see Section 3.5.3.). Also in this case, the value of the slope coefficient must be given, and then pressing the *Make Baseline* button, the background will be generated. Definitions of the Shirley functions are described in Section 3.5.3.

The *Tougaard* background described by the *Universal Cross -Section* function is the more appropriate baseline for almost all the materials because derived from theoretical description of loss processes. However, the definition of the *Tougaard* background requires a large spectral range extending at least 50 eV at higher BE from the core-hole position. This condition is generally not satisfied in common situations where a more narrower energy range around the peak is selected to acquire the high-resolution spectra resulting in a sensible reduction of the acquisition time. Similar to the *Shirley*, also for the *Tougaard* background, *2P.Tougaard* (the classic

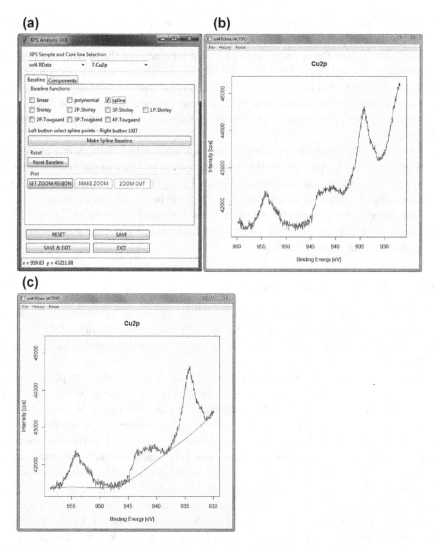

FIGURE 3.27 (a) The *Spline* background is selected. (b) Pressing the right mouse button, the user can define a set of points under the spectrum where the background will pass. (c) The *Spline* baseline is defined pressing the button *Make baseline*.

Tougaard background), *3P.Tougaard*, or *4P.Tougaard* functions are defined to better follow the spectrum at the ends (Figure 3.26a). See Section 3.5.4 for their definition and more information. Finally, in some cases none of the previously defined functions is appropriate to describe the spectrum losses. Sometimes the background presents an uncommon trend as that of the Cu 2p shown in Figure 3.27. In these cases, there is not a simple function to describe the losses. One solution is the possibility to use a polynomial function although there is not any theoretical ground supporting this choice. Selecting the *Polynomial* function, the user must enter the polynomial degree and press the button *Make Baseline* to fit the experimental data with the selected polynomial. The button *Reset Baseline* can be used for changing the polynomial degree and create a new baseline. Another possibility is to create a baseline using a *Spline* function, although also in this case no theoretical grounds support this choice. The *Spline* function may help to define a baseline in anomalous situations. The *Spline* is a curve passing through a set of points defined by the user under the spectrum as shown in Figure 3.27a.

When the *Spline* baseline is selected, the user must define these points using the right button of the mouse. In *RxpsG*, they will appear in green in the graphical window. The *Spline* points are indicated in Figure 3.27b. Then pressing the button *Make Baseline*, the two points at the extremes are chosen as boundaries of the *Region-to-Fit* and the *Spline* curve through the set of points is generated as shown in Figure 3.27c. Again the button *Reset Baseline* cancels the background and allows restarting the procedure to create a new baseline.

3.8.2.   PEAK FITTING: SELECTION OF THE CORRECT LINESHAPES

When the baseline is defined, *RxpsG* enables the choice of the fitting component lineshapes. This is done selecting the *Components* page of the *Analyze GUI*. Under *ADD/Delete*, a drop-down menu lists all the fitting functions available as illustrated in Figure 3.28.

After selection of the function, the intensity and position where to place the fit component must be defined. This is done by simply left clicking on the spectrum where the user wants to add a fit component, a cross is created in that position as shown in Figure 3.28b. Next clicks allow adding further fit components with the selected lineshape. This process is repeated as many times as the components required for fitting the spectrum as shown in this example. The component sum forming the fit function (continuous line) is updated upon addition/elimination of fit components. The fit function does not need to perfectly match the spectrum. The position and intensity of each single component can be changed by selecting the page *Move* and the component to move by the drop-down menu. Upon selection, the correspondent spectral component is indicated with a marker. Just left-clicking, the marked component is moved in the mouse position as illustrated in Figure 3.28c for the component C3.

In *RxpsG* are defined a rich number of fitting functions, including *Gaussian*, *Lorentzian*, and *Voigt*; these same functions are defined with an asymmetric tail on the high BE side. Additional lineshape are the asymmetric *Doniach-Sunjic*, and this same function with a Gaussian broadening or a tail. If needed, additional fit function can be easily implemented.

Fitting components can be removed by pressing the *Delete* button as in Figure 3.28d. A drop-down menu allows the selection of the component to delete. Be careful, removal of a fit component is possible only removing all the links on the fitting parameters which were set to fit the spectrum (e.g., links on the component position, or relations between spin-orbit component intensities or their width).

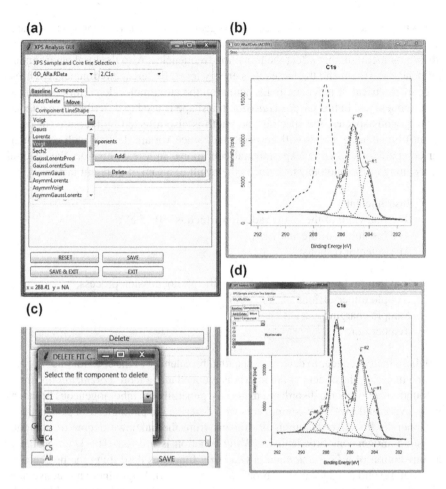

FIGURE 3.28 (a) The *Components* option page of the *Analyze* GUI. A drop-down menu lists all the fit function available in *RxpsG* and enabling the selection. (b) After pressing the *Add* button, left clicking with the mouse on the spectrum in the desired position a fit component is added. (c) Selected fit components can be adjusted under the spectrum using the *Move option*. (d) Fitting components can be removed using the *Delete* option and selecting the redundant one.

3.8.2.1. Selection of the Fit Function

As already observed in section 3.6.4, the *Lorentian* function describes the spectral lineshape associated to transitions in atoms, molecules, and solids where the width of the spectral line is related to the uncertainty on the energy ΔE and Δt with $\Delta E \Delta t \geq \hbar$. Then, $\Delta E \approx \hbar / \Delta t$ represents the minimum broadening associated to spontaneous decay and is called *natural* broadening. The spontaneous decay is described by an exponential function, and the associated line broadening affects equally the transitions of all the atoms and molecules. In this case, it is possible to demonstrate that the associated lineshape is a *Lorentzian* function.

However, there are other sources of broadening, thermal (or Doppler) broadening, as the motion of atoms or molecule rotations, the turbulence in the radiating or absorbing medium, that affect the transition frequencies. Also the local environment influences the atom and the molecule's energy level. Different environments or presence of electrical or magnetic fields acting on the atom/molecule perturb the transition frequencies. In this case, a *Gaussian* lineshape is used to describe these effects.

In gaseous species, the spectral line width is also affected by pressure (or collisional) broadening which will be neglected, since we are dealing with solid matter compatible with UHV experimental conditions. In a real situation, the spectral broadening of the photoelectron line is the result of a sum of different factors:

- Instrumental
  The linewidth of the X-ray radiation which is ~ 0.25 eV;
  The resolution of the analyzer;
  The detector response;
- Intrinsic
  The natural broadening associated to the photoemission process;
  Sample imperfections
  Sample charging
  Temperature

All these factors concur to determine the final broadening of the spectral line. In crystalline material, the effects of the matrix is reduced and a more *Lorentzian* lineshape is found. In amorphous, disordered materials, generally the inhomogeneous broadening prevails and *Gaussian* components well describe the photoelectron peaks.

Generally, to solve the ambiguity raising from the unknown degree of homogenous/inhomogeneous broadening, a *Voigt* function is utilized. The *Voigt* profile is a convolution of a *Lorentzian* and a *Gaussian* functions describing the presence of both natural and casual source of broadening. The use of *Voigt* function is convenient also because the *Lorentzian/Gaussian* mix is parameterized and optimized by the fit algorithm to better describe the actual spectrum.

Finally, here is a note regarding the *Doniach-Sunjic* function. As seen in section 3.6.4, this profile is utilized to describe asymmetric core-lines in metals and conducting materials as graphite. In these systems, the overlap between valence and conduction band creates a continuum of the excited level available in loss processes. This continuum of states is seen in the photoemission spectrum with an increase of the intensity of the photoemission peak on the high BE side. This is due to the cascade of loss processes which can occur leading to progressive reduction of the ejected electron energy, thus contributing to the spectrum at higher BE. In the *Doniach-Sunjic* function, the asymmetry factor α accounts for the intensity of the loss processes. For $\alpha = 0$, the function reduces to a simple *Lorentzian* function. In *RxpsG*, there is a collection of simplified asymmetric functions obtained by modifying the original *Lorentzian*, *Gaussian*, and *Voigt* functions. However, the more correct lineshape is the *Doniach-Sunjic* which is supported by a theoretical description of the losses by excitation of quasi free electrons. *Gaussian* broadening can also be added to this function when needed.

After definition of the baseline, selection of correct lineshape, and addition of all the fitting components needed to describe the spectrum, the *Analyze* procedure is concluded. Then pressing the *Save and Exit* button the user can proceed with setting component's constraints and best fitting.

3.8.3.   FITTING PROCEDURES

3.8.3.1. Fit Constraints

Best fitting is done by varying the fitting parameters, that is, component position, intensity, full width at half maximum (FWHM), and asymmetry, to minimize the deviation of the fitting function from the experimental data. Generally, some constraints on the variation of the fitting parameters are set before starting the fitting procedure. For example, the FWHM of the fitting components should be almost the same for all the fitting components. For a given element, the energy splitting and the relative intensity of spin-orbit components are well known. Spin-orbit positions and intensities can then be linked. It can be helpful to limit the possible energy range or fix the position of fitting components to facilitate the bond assignment and the interpretation of the spectrum. The parameters associated to each of the fitting components are controlled using the option *Fit Constraints* illustrated in Figure 3.29.

In Figure 3.30a is shown the *Parameter Table* used to set links among the fitting parameters of the list of components used to fit the C 1s reported in Figure 3.29.

To show how this matrix is composed, different fitting functions are selected to fit the C 1s. Component C1 is a *Doniach-Sunjic* function; components C2 and C3 are *Voigt* functions, while components C4, C5, and C6 are *Gaussian* lineshapes. All the fitting parameters are described in a *Parameter Table* shown in Figure 3.30a. In the upper part of the figure, there is a matrix of checkboxes. Each of the columns of the *Parameter Table* corresponds to a fitting component. In the first column are listed the fitting parameters of component C1. It has five elements corresponding to

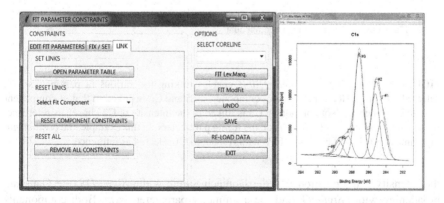

FIGURE 3.29   The *Fit Constraints* GUI used to set the constraints on the fitting parameters.

(a)

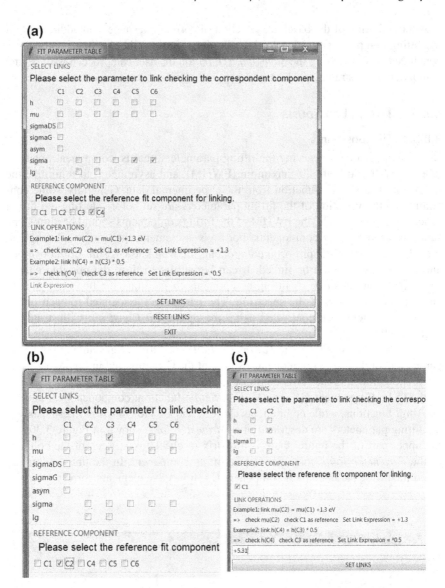

FIGURE 3.30 (a) The *Parameter Table* enables linking the various fit parameters to a selected reference. Here, the sigma of components C5 and C6 is linked to that of component C4. (b) The intensity *h* of component C3 is linked to the intensity of C2 chosen as reference. (c) It is also possible to set *relations* on the fitting parameters. In this example, the position *mu* of component C2 must be at +5.31 eV with respect to C1 selected as reference.

the intensity *h* the position *mu*, the Doniach-Sunjic width *SigmaDS*, the Gaussian broadening width *SigmaG*, and the asymmetry parameter *asym*. Both components C2 and C3 are *Voigt* and are defined by the four parameters *h*, *mu*, *sigma*, and the Lorentzian/Gaussian mix *lg*. The remaining *Gaussian* components C4, C5, and C6

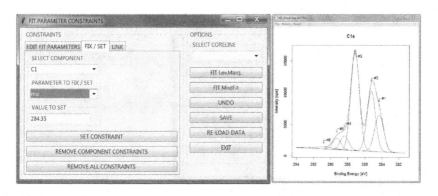

FIGURE 3.31 The *FIX/SET* page of the *Fit Constraints* GUI.

are defined just by the three parameters *h*, *mu*, and *sigma*. Linking a parameter using the *Parameter Table* is easily done checking the correspondent boxes and selecting the reference component. Note that links can be set among components of the same nature. In Figure 3.30a, the *sigma* of components C5 and C6 are linked to the *sigma* of component C4 (i.e., will assume the same value). Similar operations can be done for each of the fit parameters. The constraint is set by pressing the button *Set Links*. Observe that the *sigma* act differently in the *Voigt* or *Gaussian* lineshapes. For this reason, it is not possible to link parameters belonging to different fitting functions. In the present case, the user then can add another constraint to force the width of the *Voigt* components be the same. As shown in the example of Figure 3.30b, the *sigma* checkbox of components C3 is marked and component C2 selected as a reference. The constraint is set pressing the *Set Link* button as already noted. In this *Parameter Table* it is also possible to set specific conditions on the parameters. For example, in spin-orbit doublets, the user can link the position and the intensity of the two spin-orbit components. This is shown in the example for the Au 4f spin-orbit doublet described by two *Voigt* functions with an energy splitting $\Delta E = 5.31$ eV. To set this constraint, first check *mu of* C2, then check C1 as component reference, and finally set the *Link Expression* equal to +5.31 as shown in the example of Figure 3.30c, and press the button *Set Link*. Similar procedure is applied for the other parameters. For example, the user can check *h* of component C2 and select C1 as a reference and *Link Expression* = *0.75 to force the intensity of the two spin-orbit components to respect the 3:4 ratio (see Section 3.5.2).

The second page of the *Fit Constraints* permits to set/fix the value for each of the fitting parameters. As an example, the graphitic phase of the C1s spectrum shown in Figure 3.31 is assigned to component C1. Graphite and graphitic phases fall at 284.35 eV.

Using the *Parameter to Fix/Set* drop-down menus, the user can select both the fit component and the fit parameter to fix, which in the example of Figure 3.31 are C1 and *mu*, respectively. Then 284.35 eV is assigned to the *Value to Set* editable row. Finally the *Set Constraint* button will force *mu* to be fixed at the given value. Same procedure is applied to the other fitting parameters. Figure 3.32 illustrates the *Edit Fit Parameters* page of the *Fit Constraints* GUI. Upon selection of the desired fit

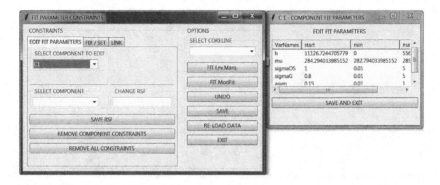

FIGURE 3.32   The *Edit Fit Paramater* page of the *Constraints* GUI and the window listing the fitting parameters which can be edited.

component, a panel containing all the correspondent fitting parameter is opened. For each of the parameters, the user can modify the values and the default range of variability which will be utilized by the fitting algorithm. In Figure 3.32 is selected the first C1 component which corresponds to a *Doniach-Sunjic-Gauss* function, on the right are shown the correspondent parameters listed in the editable panel. As an example, the user can limit the variability range of *SigmaDS* changing the *min*, *max* values to 0.3 and 1, respectively. The *Save and Exit* button will update the values of the fitting parameters and closes the edit window. Similar procedure applies for each of the fit components of the selected core-line.

In this page, it is also possible to change the *Relative Sensitivity Factor* of each individual fit component just selecting it and giving the new RSF value. This is also possible in the *Quantification* procedure.

As the constraints have been set, the *Save* button on the right of the *Fit Constraints* GUI must be pressed (see Figure 3.29). This starts a control on the parameters, if they are correctly defined. *Levenberg-Marquardt* and *Model Fitting* options are available to perform the peak fitting. In general, *Levenberg-Marquardt* is selected while *Model Fitting* is chosen when the first option fails. *Model Fitting* offers a list of different algorithms to perform peak fitting and will be described in the following. Generally, the *Levenberg-Marquardt* algorithm is fast and stable. It may happen that a *convergence error* appears when the best fit is obtained with one less component. The best fit generated and the relative fitting parameters are saved and available. The user can modify the final value of the component intensity and the variability range to avoid the fit component be set to zero, for example, limiting the width of the other components. The *Save* button must be pressed to retain the best fit results. The button *Undo* restores the initial conditions to restart the fit procedure with different parameter constraints and variability ranges. The button *Re-Load Data* is used in combination with the *Move Components* option and will be described in the following.

3.8.3.2.   Best Fit

The two options available in *RxpsG* to activate the algorithms for best fitting were already mentioned in the *Fit Constraints* GUI. They consist in two different approaches namely the *Levenberg-Marquardt* algorithm and the *Model Fitting*. The *Levenberg-Marquardt* algorithm was discovered first by Levenberg K. in 1944 [76] and rediscovered by Marquardt in 1963 [77] and is based on the damped least-squares method and used to solve non-linear least squares problems.

Linear least squares refer to a system of linear equations representing the differences between experimental values and their corresponding modeled values. The best approximation is the one that minimizes the sum of squared differences and the system of linear equations has a unique solution. However, not always the differences may be described by linear equations. *Non-Linear Least Squares* is used to fit a set of m experimental data with a model that is non-linear in n unknown parameters ($m \geq n$). The basis of the method is to approximate the model by a linear one and to refine the parameters by successive iterations. In R, the *Non-Linear Least Squares* algorithm is provided by an interface which allows for defining upper and lower parameter bounds. These are the values that the user can freely adapt using the *Fit Constraints* GUI by editing the fitting parameters associated to each of the fitting components.

The Levenberg-Marquardt method combines the Gauss-Newton with a gradient descendent algorithm. The Gauss-Newton algorithm assumes that locally the difference between experimental data and model is linear. However, if the guess is too far, the Gauss-Newton algorithm fails while the LM approach converges to the correct solution.

As the fitting components are defined and the constraints are set, the best fit is immediately obtained by selecting the *Levenberg-Marquardt* option from the *Analysis* menu or pressing the relative button in the *Fit Constraints* or *Move Component* GUIs. Upon best fitting, sometimes a *non-convergence* warning is generated. The error the most of time is generated because one of the fitting components is null and the best fit is done without using that component. *Fit Constraints* can be used to include that fit component in the best fit if this is needed to be in agreement with the chemical interpretation hypothesized for that material.

3.8.3.3.   Model Fitting

The problem with gradient descent algorithms, namely the *Levenberg-Marquardt* algorithm, is that it will converge into a minimum that is nearest to the initial guess, so when starting from different positions, you will end up in different local minima. In this case, it is possible to refit the same spectrum using the *ModFit* interface which is more robust. The *ModFit* interface displayed in Figure 3.33 allows using also other fitting algorithms as shown in the window. These algorithms are grouped in three different classes:

- *Classic algorithms* include *Levenberg-Marquardt, Gauss-Newton,* and *Port* algorithms;
- *Conjugate-gradient algorithms* encompass the *Nelder-Mead, Conjugate Gradient, BFGS,* and *L-BFGS-B* algorithms;
- *Random algorithms* are the *Sann* and the *Pseudo-Random* algorithms.

The interface is self-explaining. When the background and the fitting components are defined, the user can freely select the fitting algorithm which is activated by

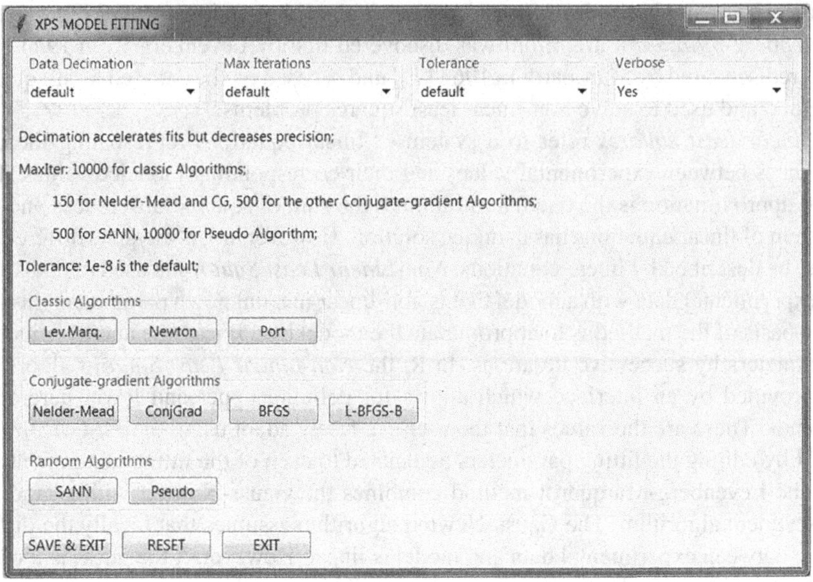

FIGURE 3.33   The *Mod-Fit* interface.

pressing the relative button. For each of the algorithm are defined default values for the *Decimation*, *Max Iterations*, *Tolerance*, and *Verbose*.

Decimation is set to 1, that is, no decimation is applied, for all the algorithms. However, random algorithms are rather slow. Depending on the number of data, in this case, a *Decimation* = 2 or 3 may help to speed up the fitting procedure.

Max-Iterations is the number of maximum iteration allowed to fit the data. For *Levenberg-Marquardt* and *Newton*, it is set to 10,000. For *Port*, *Nelder-Mead*, *CG*, *BFGS*, *L-BFGS-B*, *SANN* *MaxIteration* = 200, for *Pseudo*, it is 500. The default *Tolerance* = 10^{-9} and *verbose* is set to *Yes*. Indications are given to correctly set these parameters in relation to the algorithm selected. As for the fitting procedures, there are substantial differences among the various algorithms. A brief description of the algorithms is now provided to help the user in selecting the correct fitting method.

3.8.3.3.1   Classic Algor4ithms

Gauss Newton   The *Gauss Newton (G-N)* algorithm [78, 79] is used to minimize the non-linear least squares defined by

$$S(x) = \tfrac{1}{2}\Sigma_j\left[y_j - p(x_j)\right]^2 = S = \tfrac{1}{2}\Sigma_j F(x)^2 \qquad 3.36$$

where j = 1... m are the experimental data, y_j is the generic experimental value, and $p(x_j)$ is the value of the correspondent function used to model the y data. The problem is the minimization of the sum of least squares S. The most solution is to compute derivatives to find the minimum of the S function.

$$S'(x) = \Sigma_j F_j(x) F'(x) \qquad\qquad 3.37$$

Introducing the Jacobian $J(x) = F'(x)$, the gradient of the least squares may be written in matricial form as

$$\nabla S(x) = J^T F \qquad\qquad 3.38$$

If the residuals $F(x_j)$ are small (corresponding to a good initial estimate), the *G-N* has a simplified expression, and the S(x) may be obtained by iteration

$$x_{n+1} = x_n - (J^T J)^{-1} J^T F(x_n) \qquad\qquad 3.39$$

Levenberg-Marquardt   The *Gradient Descendent* used in the *Levenberg-Marquardt* (*L-M*) method introduces a term in the *G-N* algorithm. The iteration scheme is now

$$x_{n+1} = x_n - \lambda_n \nabla S(x) \qquad\qquad 3.40$$

Then the *L-M* approach [80] to minimize the leas-squares becomes

$$x_{n+1}(\lambda) = x_n(\lambda) - (J^T J + \lambda I)^{-1} J^T F(x_n) \qquad\qquad 3.41$$

where I represents the identity matrix. The parameter λ is adjusted at each iteration. If the $\nabla S(x)$ decreases rapidly, a small value of λ is used and the algorithm behaves as the *G-N*. In the opposite case, when the variation of $S(x)$ is small, λ is increased and the variation of x_{n+1} is in the opposite direction to the gradient.

The *L-M* algorithm is the same as that described in the previous section. However, in *ModFit*, the method used to apply the algorithm is different and more robust. This appears also by the longer time required to arrive to the convergence and to the best fit.

Port   The *Port* algorithm refers to the use of the PORT3 library of routines by AT&T-Bell labs [81–84]. The PORT3 is a library covering the following mathematical areas: general minimization, nonlinear least squares, separable nonlinear least squares, linear inequalities linear programming, and quadratic programming. The nonlinear optimizers have unconstrained and bound-constrained variants.

The non-linear optimizer uses sets of starting values for the model functions. The Gradients and Jacobians can be approximated by finite differences. The minimization of the least-squares adaptively switches between the *G-N* Hessian approximation and an "augmented" approximation using a quasi-Newton update. Unless the $S(x)$ function is convex, the PORT3 optimization routines may only find a local minimum, even when "better" minima exist. This can require trying minimization with different starting points.

3.8.3.3.2 Conjugate-Gradient Algorithms

Nelder-Mead   *Nelder-Mead* (NM) algorithm [85] is a numerical method commonly used to find the minimum (or maximum) of an objective function in a multidimensional space. The objective function may represent the deviation of the model data function from the experimental depending on n fitting parameters. The NM algorithm directly search by function comparison and is often applied to nonlinear optimization problems without computing derivatives. However, this approach is heuristic and can converge to non-stationary points [86]. Essentially, given a function f of n variables x, the NM method evaluates a function in $n + 1$ points.

These can be seen as the vertex of a convex $n + 1$ dimensional "simplex" S. For example, for $f(x_1, x_2)$, the simplex is a triangle with vertexes $x1$, $x2$, and $x3$. If f depends on three variables, the simplex is a tetrahedron, etc. The NM algorithm performs a sequence of transformations of the working simplex S designed to decrease the values of f at its vertices. The transformation (reflection, contraction, expansion, shrink) consisting in determining one the position of one test points where to evaluate f. On the basis of the value assumed by f in this point, one of the transformations is applied resulting in the minimization of the objective function f. The process is terminated when the standard deviation of f with respect to the current simplex is lower than a preset tolerance threshold.

Conjugated Gradient   The *Conjugated Gradient* (CG) [87] algorithm is a numerical method to solve a set of n linear equations. Consider a linear equation $Ax = b$ where A is a $n \times n$ symmetric $\left(A^T = A\right)$ positive matrix and x and b are n-dimensional real vectors. The solution x of the set of n equations is equivalent to a minimization problem of a convex function $f(x)$ defined by

$$f(x) = \tfrac{1}{2} x^T A\, x - b^T x \qquad\qquad 3.42$$

The problem is solved iteratively defining the residuals $r(x)$

$$r(x) = \nabla f(x) = Ax - b \qquad\qquad 3.43$$

namely for the generic k element $r_k(x) = A x_k - b$. Minimization of $f(x)$ is made along the direction $x_k + \alpha p_k$ defined by the condition that $f'(x_k + \alpha p_k) = 0$.

$$\alpha_k = -r_k^T p_k / p_k^T A p_k \qquad\qquad 3.44$$

The set of $\{p_0, p_1, \ldots, p_n\}$ is selected to be conjugates with respect to matrix A, that is

$$p_k^T A\, p_h = 0 \text{ for all } k \neq h \qquad\qquad 3.45$$

so giving the name to the algorithm. The direction p_k may be selected to be the steepest descendent direction of f corresponding to r_k, that is

$$p_k = -r_k + \beta_k p_{k-1} \qquad\qquad 3.46$$

since $k-1 \neq k$ applying equation (3.45), it is possible to obtain the value of β_k

$$\beta_k = r_k^T A\, p_{k-1} / p_{k-1}^T A\, p_{k-1} \qquad\qquad 3.47$$

then knowing p_k from equation (3.46) and α_k from equation (3.44), it is possible to calculate the direction of minimization $x_k + \alpha p_k$.

BFGS   A quasi-Newton method was developed by Charles Broyden [88, 89] as an alternative to Newton's method .for solving nonlinear algebraic systems. As seen earlier, the problem of finding a zero of a nonlinear function g(x) may be solved using the Newton–Raphson starting from x_1, a suitable approximation of a zero, and continuing by iteration where $x_{n+1} = x_n - \left(J^T J\right)^{-1} J^T F(x_n)$ (see equation 3.39). The algorithm ensures convergence if the starting position x_1 is sufficiently close to a solution. In the *BFGS* method, instead of working with the true inverse Jacobian J, a suitable approximation H_n is computed. Thus, the iteration process is of the form

$$x_{n+1} = x_n - H_n F(x_n) \qquad\qquad 3.48$$

The *BFGS* determines the descent direction by preconditioning the gradient with curvature information.

Applying a Taylor expansion truncated at first term, one gets

$$g_{n+1} = g_n + J_n(x_{n+1} - x_n) \qquad\qquad 3.49$$

where g_n represents the function $g(x_n)$. Introducing $s_n = x_{n+1} - x_n$ and $y_n = g_{n+1} - g_n$, equation (3.49) may be rewritten as

$$J_n^{-1} y_n = s_n \qquad\qquad 3.50$$

Equation (3.50) describes a system of n linear equations in n^2 variables which are the components of the approximate Jacobian of g_n. Therefore, it is an underdetermined linear system with an infinite number of solutions. The problem is solved by limiting H_n to be of the form $H_{n+1} = H_n - u_n v_n^T$ and applying the so-called first update (or Broyden's good method), with $v_n = H_n^T s_n$

$$H_{n+1} = H_n - (H_n y_n - s_n) v_n / y_n v_n \qquad\qquad 3.51$$

Since *BFGS* does not compute the Jacobian and its inversion, the computations are easier and faster than those of the Newton algorithm.

L-BFGS-B   L-BFGS-B is an optimization algorithm of the *BFGS* methods using a limited amount of computer memory. The L-BFGS-B algorithm introduces upper/lower bounds for the x_i variables lower and upper bounds values. Using a simple gradient method, the method identifies fixed and free variables at every step and then applies the *BFGS* algorithm on the free variables only to obtain higher accuracy. Then apply iteration.

3.8.3.3.3. Random Algorithms

SANN   The Simulated-Annealing "SANN" method by Belisle [90] belongs to the class of stochastic global optimization methods. It is a metaheuristic method to approximate global optimization using only function values. It will also work for non-differentiable functions. It is often used when the search space is discrete, for problems where finding an approximate global optimum is more important than finding a precise local optimum in a fixed amount of time.

The problems solved by SA are currently formulated by an objective function of many variables, subject to several constraints. The performance of search algorithms as SANN relies crucially on their parameterizations, for example, the selection of correct factor settings. This implementation uses the Metropolis function for the acceptance probability and the algorithm is relatively slow.

Essentially, in the SANN algorithm, the solution is searched in the space of the solutions and checking the probability of accepting worse solutions. Accepting worse solutions allows for a more extensive search for the global optimal solution. The algorithms at each time step randomly selects a solution at point y close to the current state in x, measures its quality, and moves to it according to the probabilities of selecting better or worse solutions. In the Metropolis criterion, the probability is described by

$$P(x, y, t) = \begin{cases} \exp\left[-\left(f(y)-f(x)\right)/t\right] & \text{if } f(y) > f(x) \\ 1 & \text{if } f(y) \le f(x) \end{cases} \qquad 3.52$$

Pseudo-Random   A common algorithm that belongs to stochastic methods is the controlled random search (CRS) method [91, 92]. This is a procedure based on the use of a population of trial solutions. Initially, a set with randomly selected points is created and then the algorithm repeatedly replaces the worst point with a randomly generated point. This process iterates until some termination criterion is satisfied. Also *Pseudo-Random* methods do not require the function to be differentiable or variables to be continuous and are applied in presence of constraints. Initially, a search domain V is defined by setting the limits of the n variables and the number N of the data-points P randomly selected over V. The function $f(x_1 ... x_n) = f(P)$ is evaluated at each trial and the position P and the correspondent value of f are stored. At each iteration, a new trial point P_{n+1} is randomly selected among the N set of random points initially defined in V. $f(P_{n+1})$ is compared with $f(P_n)$. If $f(P_{n+1}) < f(P_n)$ then P_n is replaced by P_{n+1}. If $f(P_{n+1}) > f(P_n)$ the trial point P_{n+1} is discarded and a new trial point P is evaluated. The probability to converge to the global minimum strictly depends on the number of trials N, the complexity of the function, the kind of constraints and the rule adopted to select the N trial data-points P.

Similar to the *SANN*, also the *Pseudo-Random* algorithm is slow. To speed-up the fitting procedure a decimation, the number of iterations and the tolerance may be set as follows:

Decimation=Yes with decimation level consistent number of data to fit (decimation level 3 means 1 data over three is considered for the fit), *MaxIteration=500(Sann) 10000(Pseudo), Tolerance=1e-8*

3.8.4. CORE-LINE PROCESSING

The *Core-Line Processing* option is intended for manipulation of the spectrum. This option includes a list of operations which can be applied to the original spectrum which will be described in detail in the following. Frequently occurs that a series of similar samples must be analyzed. For example, a number of samples are synthesized varying the concentration of one or more elements; the same samples undergo a treatment (chemical, plasma, etc.) or thermal treatments at different temperatures, so that the composing elements are the same, but their concentration changes. In this case, the *Core-Line Processing* option offers the possibility to replicate the analysis made on the first sample to the other specimens. An example is shown in Figure 3.34a, b, where two different specimens based on GO are illustrated. Observe the different lineshape of the C 1s and O 1s and the different core-line intensities in the two samples. It is supposed that the energy scale of the source and destination *XPS-Samples* are aligned to a common reference. If charging effects are present, the *Energy Shift* must be used to get the appropriate energy scale (see Sections 3.2 and 3.3).

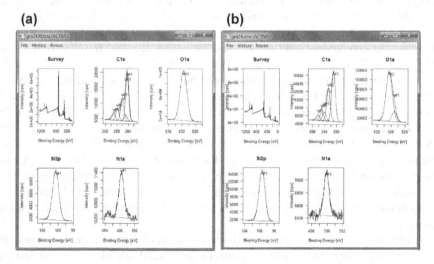

FIGURE 3.34 (a) Prototype of analyzed *XPS-Sample*. (b) Spectra acquired on a similar specimen. The analysis shows the same core-lines. Variation of the intensity and lineshapes can be present (see, for example, the C 1s).

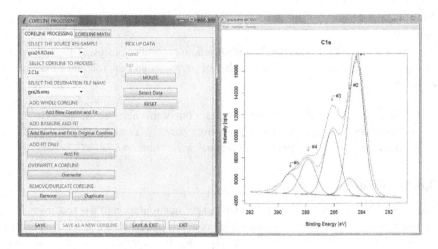

FIGURE 3.35   First page of the *Processing Core-Line* GUI.

To replicate the analysis, the user must select the prototype *XPS-Sample* as source. In Figure 3.35, this corresponds to the Gra24.Rdata file. Upon selection, the content of the prototype *XPS-Sample* is shown. Then, let us suppose to replicate the analysis made on the C 1s on the destination *XPS-Sample* Gra26.vms as shown in the example of Figure 3.35. The user must then select the C 1s as core-line to process. This core-line will be visualized (see Figure 3.35).

Finally, the user must select the destination filename. Again, upon selection, the content of the selected *XPS-Sample* will be visualized to make sure the user that the operation will be applied to the correct XPS Sample. At this point, there are three possible options: (i) *Add new Core-Line and Fit*. This option will add the entire C 1s core-line of the source *XPS-Sample* to the selected destination. If a fit is present, baseline, fit components, and fit will also be added. (ii) *Add Baseline and Fit to the Original Core-Line*. This option adapts the baseline and the fit of the selected "source" core-line to the "destination" core-line. A warning is raised if the selected source core-line is not found in the destination *XPS-Sample*. First, the software detects which kind of baseline was utilized in the source core-line. Then the user is asked to define the edges of the baseline on the destination core-line. This will start the replication of the baseline and fit. The operation is concluded pressing the button *SAVE* to store the analysis in the destination *XPS-Sample*. This sequence of operations can be repeated for the other core-lines. (iii) *Add fit*, this option permits to replicate only the fit. This operation can be applied when different baselines are used in the source and destination spectra because of the different loss contributions in the two cases.

The next option *Overwrite* regards the replacement of a destination core-line with the correspondent in the source. Just select source and destination files, the desired core-line, and press *Overwrite* and then *SAVE*. The next options regard the duplication or the elimination of a core-line in an *XPS-Sample*. In this case, source and destination files must be the same, pressing *Duplicate* the *Core-line to be processed*

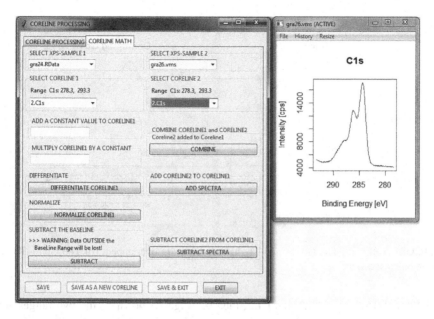

FIGURE 3.36   The second page of the *Core-Line Processing* option.

will be duplicated while pressing *Remove* it will be eliminated. The operation must be concluded by pressing *SAVE* to store the results of the operation in the destination file.

The last option regards the possibility to clip part of a spectrum where no baseline or fit are present. The user can define the extension of the region to extract entering the *from* and *to* values or defining the region to extract using the mouse in a mode similar to that described in Section 3.4.3. The operation is completed by pressing the button *Save as a new Core-Line* to store the culled data in a new core-line.

The second page of the *Core-Line Processing* shown in Figure 3.36 is dedicated to math operations on spectra of the single *XPS-Sample* or on spectra from two different *XPS-Samples*. At the top part of the GUI, the user can select the *XPS-Sample1*, and the core-line CL1, the *XPS-Sample2*, and the core-line CL2. Upon selection of the core-lines, the correspondent x, y ranges are visualized.

In the remaining part of the GUI, various mathematical operations can be applied on a single or on a couple of spectra. The user can add a constant value to the CL1, or multiply CL1 for a constant value by entering the relative value and pressing *ENTER*. It is also possible to differentiate or normalize the CL1 spectrum just pressing the correspondent buttons. Computing CL1 + CL2 or CL1 − CL2 is obtained pressing the *Add Spectra* or the *Subtract Spectra* buttons (observe that subtraction is always done in the indicated core-line order). Finally the last option *Combine* offers the possibility to merge CL1 and CL2 spectra. It may happen that CL1 and CL2 are overlapped as shown in the example of Figure 3.37a.

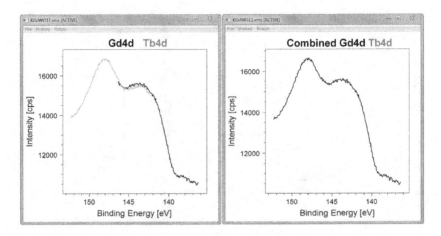

FIGURE 3.37   (a) The Gd 4d and Tb 4d core-lines are overlapped. (b) The Gd4d–Tb 4d combined spectrum.

Upon selection of CL1 and CL2, their X-range is shown. If the two ranges are overlapped, CL1 and CL2 can be combined. The software automatically computes the extension of the overlap and the intensity of the two core-lines and proceeds to a leveling then joins the spectra. The result of the sequence of these operations is shown in Figure 3.37b.

3.8.5. ADJUST BASELINE

Sometimes it is required to redefine the extremes of the *Region to Fit* or the level of the *Baseline*. The *Adjust Baseline* function is made to solve these problems. Figure 3.38 shows the simple interface of this option.

Once the XPS Sample and the desired core-line are selected, the user can just press the button *Set Baseline Boundaries* to define the new limits/level of the baseline (i.e., the new extension of the *Region to Fit*) as shown in Figure 3.38. Just left click on the spectrum with the mouse to define the baseline ends. The definition of the baseline edges will result in an automatic update of the baseline and of the *Region to Fit*. If a fit is present, also the fitting component and the fit will be changed consistently with the new baseline.

A zoom option is also available to facilitate the definition of the baseline ends. The selection of the zooming area is made with the mouse defining two opposite corners of the zooming area similar to what described previously. *SAVE* and *SAVE & EXIT* conclude the procedure.

3.8.6. SMOOTHING

Spectral analysis frequently requires a high value of signal-to-noise ratio (SNR). This, for example, occurs when differentiation of the spectra are required. In this

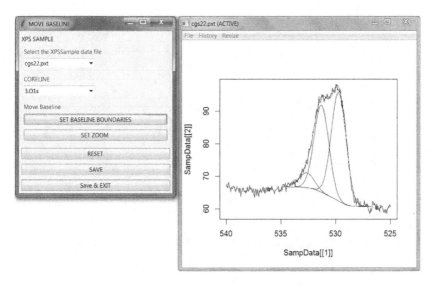

FIGURE 3.38 The *Adjust Baseline* interface.

specific case, the digital first derivative corresponds to the difference of adjacent data. The result of this operation is maintaining the high-frequency noise, while the lower frequencies characterizing the XPS spectrum are reduced (see Figure 3.39a). In these circumstances, the solution is to remove the noise by filtering the signal to obtain a meaningful differentiated signal as in Figure 3.39b. The *Smoothing* option illustrated in Figure 3.39c is intended for offering a list of filters for noise removal.

For each filter, the user can select the filter order, the degree of noise rejection and the option to subtract a linear background to reduce its effect on the smoothing procedure. Increasing the filter order corresponds to a better noise removal (higher steepness of the filter transfer function). Let us enter in a more detailed description of the various filters. Observe that the characteristics of a filter are visualized by the transfer function which represents the frequency response of the filter and is obtained as the filter response to a step function. This option is also provided by *RxpsG*. The step function is used because the transition from 1 to 0 value in a null interval of energy contains all the frequencies and then the response of a filter represents its capability of noise rejection, presence of oscillations, etc.

3.8.6.1. Filtering: Energy and Frequency Domains

XPS spectra are representation of the intensity of the photoelectron signal as a function of the energy (kinetic or binding energy) while filters are described in the frequency domain with apparent no relation with the energy domain. To understand how to select the correct filter settings (degree of noise rejection and the number of filter

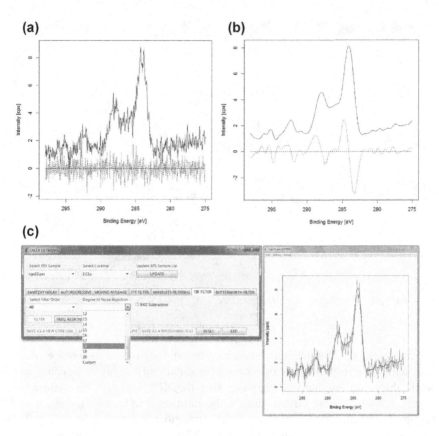

FIGURE 3.39 (a) First derivative (dashed line) of the original signal (solid line) and (b) First derivative of the same data filtered using a 40 coefficient FIR filter with noise rejection degree = 18 (solid line) and the correspondent first derivative (dashed line). (c) The *Smoothing* GUI and effect of the FIR filter applied (solid line) on the original data (dashed line).

coefficients) and correctly smooth the original spectrum, it is important to understand what frequencies have to do with XPS spectra. Since the signal is described as a function of the energy, the first question to answer is how to correlate the energy scale to the frequencies which will be retained or rejected by the filter applied to clean the spectrum. The answer is actually rather simple and is related to the conditions set for the acquisition of the spectra. Besides the energy step defining the resolution which will be used, the settings include the integration time for any channel. The integration time essentially defines the "sampling frequency" used to acquire the spectra. If the integration time is a second, the energy will be scanned with a speed of dE/second where dE is the energy step set for the acquisition. In other words, a point of the spectrum is acquired every second. As a consequence of this, it is possible to visualize the XPS spectrum as a time-sequence of values. Any time sequence can be described in the frequency domain, for example, applying a fast Fourier transform. Now we can readily select the optimal filtering condition by looking at the frequency response of the filter. A filter is characterized by the frequencies left which are described by the extension of the *pass-band* and by frequencies which are rejected representing the *stop-band*. The pass-band and stop-band will extend till to 0.5 of the sampling frequency. This is a consequence of the Nyquist theorem known as the sampling theorem. It states that, to accurately reproduce a pure sine wave measurement, the sampling rate must be at least twice its frequency. Then if we have a sampling frequency of one hertz we cannot describe frequencies higher than 0.5 Hz as it appears in the frequency response of the filter. Important in this response is also the degree of rejection, which is measured in dB and describes the power of the filter to block the frequencies in the stop-band. An ideal filter possesses a step-like frequency response equal to 1 in the pass-band which corresponds to an admission of all the frequencies in the pass-band range without distortions or amplification/attenuation till to the cut-off frequency. Frequencies higher than the cut-off frequency are blocked with a high degree of rejection corresponding to the negative dB value. However, a real filter behaves differently, the transition from the pass-band to the stop-band is not step-like but it shows a certain slope to reach a certain degree of rejection. This corresponds to a smoothed block of the undesired frequencies. Increasing the filter power generally increases the steepness of the pass-to-stop band transition and the extent of the filter noise rejection namely frequency attenuation. The shape of filter frequency response is formed by some lobes: the first is the pass-band with the highest positive intensity, the other lobes are negative indicating that the correspondent frequencies are blocked. In the next sections, examples of frequency responses of the various filters with selected degree of noise rejection and number of coefficients are shown.

3.8.6.2. Sawitzky Golay

Savitzky-Golay (SG) is a digital filter performing a smoothing by a local polynomial regression on a series of equally spaced values. This is obtained by a convolution of successive sub-sets of adjacent data points with a low-degree polynomial [93, 94].

$$P(n) = \Sigma_k a_k n^k \qquad\qquad 3.53$$

with $k = 0, ..., N$. For equally spaced data, it is possible to find a set of "convolution coefficients" satisfying the least square equations

$$\varepsilon_N = \Sigma_n [P(n) - x(n)]^2 = \Sigma_n (\Sigma_k a_k n^k - x(n))^2 \qquad\qquad 3.54$$

where n runs on the 2M+1 data of the $x(n)$ input sequence. To find the SG coefficients, the errors ε_N are minimized with respect to a_i, that is, $\partial \varepsilon_N / \partial a_i$. This yields to a set of N+1 equations with N+1 unknown a_i. The coefficients can be applied to all the sub-sets to give estimates applying the linear least squares. When the data points are equally spaced, an analytical solution to the least-squares equations can be found, in the form of a single set of "convolution coefficients". These coefficients can be applied to all data sub-sets to produce estimates of the smoothed signal. Increasing the degree of the filter increases the number of the convolution coefficient and the degree of smoothing. The filter is rather efficient although it presents some

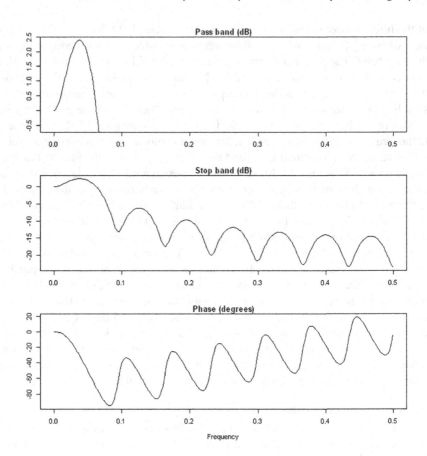

FIGURE 3.40   Frequency response of the SG filter with a degree of noise rejection = 10.

drawbacks as described in [95]. The frequency response of the SG filter is illustrated in Figure 3.40. As anticipated before, the frequency response tells us the behavior of the filter in the frequency domain. We see that in the pass-band the filter response is not flat but it shows a parabolic shape with an average value around 1. This tells us that some frequencies are admitted with higher extent with respect to very low frequencies and frequencies near to the cut-off frequency located at ~0.1 Hz for an SG filter with degree of rejection equal to 10. The stop-band is formed by a sequence of lobes with an oscillating degree of rejection of about −7 dB in the first lobe and slightly lower rejection of the next lobes. Then the SG filter is a rather mild filter with limited noise rejection.

3.8.6.3.   Autoregressive Moving Average

Autoregressive Moving Average models were introduced in past middle of 19th century to describe time series in the case of weakly stationary stochastic processes

[96]. The model is composed of two polynomials. The *Autoregressive* (AR) model is based on a polynomial formed by past terms of the time series. The *Moving Average* (MA) model is formed by a linear combination of previous and successive terms with respect to the actual time t. Both models may be applied to describe the time series and in this case the model is referred to as ARMA(p, q) where p and q are the orders of the AR and MA parts respectively. The mathematical expression of the AR filter response of order p is [97]

$$Y_t = c + \Sigma_j a_j Y_{t-j} + \varepsilon_t \qquad 3.55$$

where a_j with $j = 1,...,p$ are the coefficients of the AR model, c is a constant, and ε_t represents a white noise. In an AR model, the effect of previous time spots depends on the relative coefficients at that particular period of time. The value at time t is the result of the regression of past time series, and for this reason, the model has an infinite response, because Y_{t-j} values will affect infinitely far into the future Y_t, see equation (3.55). However, Y_t will affect Y_{t+1}, and so on forever.

In the case of an MA model, the value at time period t is impacted by the unexpected external factors at $t-1$, $t-2$, $t-3$, ..., $t-k$. These unexpected factors represent *Errors* or *Residuals* ε_t. Also for the MA, the impact of previous time spots depends on the value of coefficients b_t at that particular period of time.

For the MA model of order q, the expression is

$$X_t = \mu + \varepsilon_t + \Sigma_j b_j \varepsilon_{t-j} \qquad 3.56$$

where b_1, ..., b_q are the MA coefficients, μ is the expectation of X_t, and ε_t, ε_{t-j}, ..., ε_{t-q} represent the white noise. In the MA model, ε terms affect X values only for the actual period for a finite number of q periods into the future.

The implementation of AR and MA models as filters is straightforward. For an AR filter, the response Y_n is given by

$$Y_n = a_0 X_n + \Sigma_j a_j Y_{n-j} \qquad 3.57$$

being X_n the nth value of the original data. As observed earlier, AR filters behave as infinite impulse response filters (IIR). In the case of MA, the filter is described by:

$$Y_n = \Sigma_j b_j X_{n-j} \qquad 3.58$$

The MA filter is inherently stable, since the response Y_n depends on the value of the previous q signal values as in a finite impulse response (FIR) filter. The higher the number of coefficients p, q, the higher the noise rejection obtained.

The characteristics of the AR and of the MA filters for noise rejection degree = 10 are reported in Figure 3.41a, b. As it can be seen, the response of the two filters is similar. In the pass-band, both the filters do not show a flat response and similar is also the degree of rejection in the stop-band.

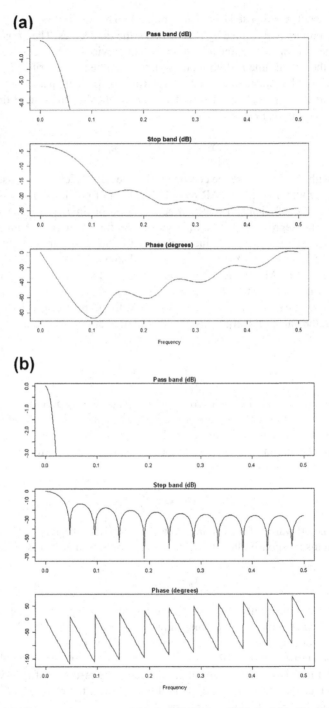

FIGURE 3.41 Frequency response of (a) the AR filter and (b) the MA filter for degree of noise rejection = 10.

3.8.6.4. FFT Filter

Digital filtering techniques essentially remove part of the frequency spectrum of a given signal which can be high-frequency noise (low-pass filter), removing the unwanted low frequencies superposed to high-frequency information (high-pass) or just select a range of frequencies containing the information (band-pass filters). In all the cases, rejection of a selected part of the spectrum can be performed in the frequency domain by applying a Fast Fourier Transform (FFT) to the signal. The frequency signal is then multiplied by a filter function which rejects the unwanted frequency components. The resulting frequency signal is then converted back to the time domain via an inverse FFT transform (IFFT). The advent of an efficient algorithm to compute the FFT [98] had deep implications for the implementation and diffusion of many signal processing algorithms. Digital signals are characterized by a series of data as a function of a physical variable, the energy on our case. The signal is then represented by a sequence x(n) and the Fourier transform is defined as

$$X(e^{i\omega}) = \Sigma_n x(n) e^{-i\omega n} \qquad 3.59$$

where $n = -\infty, ..., \infty$ and its inverse transform

$$x(n) = 1/2\pi \int_{-\pi}^{\pi} X(e^{i\omega}) e^{i\omega n} d\omega \qquad 3.60$$

For discrete sequences of time series, equations (3.57) and (3.58) become

$$X(e^{i\omega}) = \Sigma_k x(k) e^{-i2\pi/k} \qquad 3.61$$

Where $n = 0, ..., N-1$ and its inverse transform

$$x(n) = 1/N\Sigma_k X(k) e^{i2\pi/k} \qquad 3.62$$

The FFT filter is conceptually very simple. The input signal is transformed in the frequency domain via the FFT. Here, a portion of the high-frequency spectrum is set to zero on the basis of the degree of the noise rejection selected. Then a filtered signal is obtained applying the inverse of the FFT transform. To characterize the filter, we rely on equation (3.59). In that equation, $X(k)$ representing the portion of the spectrum rejected represents the coefficient of the FFT filter. Using these, coefficient is possible to calculate the filter frequency response shown in Figure 3.42 for a degree of noise rejection = 5.

3.8.6.5. Wavelet Filter

Similar to the FFT filter, the *Wavelet* filter is also based on a time-to-frequency transform made using 'wavelet' functions. Instead of exponential functions, the *continuous wavelet transform* (CWT) is based on a set of base functions, the wavelets [99]. For a sequence of data $x(t)$ in the time domain, the CWT is defined by

$$X_w(s,\tau) = 1/\sqrt{s} \int_{-\infty}^{\infty} x(t) \psi^*[(t-\tau)/s] dt \qquad 3.63$$

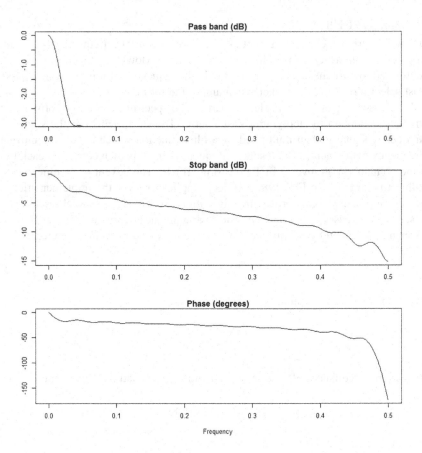

FIGURE 3.42   Frequency response of an FFT filter for degree of noise rejection = 5.

Similar to the continuous FFT, the CWT is also redundant if it is evaluated on the possible vales of t and s. For this reason, the CWTs are usually computed on a discrete grid on the time-frequency and timescale plane, which corresponds to a discrete set of continuous basis functions, and to a discrete wavelet transform (DWT). Using a discrete grid, the integral in equation (3.60) will be replaced by a summation over a discrete number of t intervals.

There are different kinds of base wavelets: Haar, Morlet, Addison, Hermitian, Daubechies, Coiflet, and Beylkin. Wavelets are generated by starting from a "mother wavelet" and applying a dilatation. For example, the mother Haar wavelet is defined as

$$\psi(t) = \begin{cases} -1/\sqrt{2} \text{ for } -1 < t \le 0 \\ 1/\sqrt{2} \text{ for } \quad 0 < t \le 1 \\ 0 \text{ otherwise} \end{cases} \qquad 3.64$$

By applying dilatation, the set of $\psi_{j,k}(x)$ base orthonormal wavelets can be defined by

$$\psi_{j,k}(t) = 2^{j/2}\psi(2^j t - \tau) \qquad 3.65$$

Similar to the FFT, given a function f(x), we can decompose it into the following generalized series as

$$f(t) = \Sigma_j \Sigma_k d_{j,k} \psi_{j,k}(t) \qquad 3.66$$

where j and k are integers with j, k $= -\infty, ..., +\infty$, and $d_{j,k}$ are called the wavelet coefficients of $f(t)$ and correspond to the factor multiplying $x(t)$ in equation (3.60) where the integral becomes a summation.

There are substantial differences between the FFT and the WT and the principal is that the WT transform was introduced to analyze non-stationary signals. The wavelet transform will provide not only information concerning the frequencies characterizing the signal but also when frequency changes occur. To do this, the input signal is windowed with decreasing window amplitude. The narrower the window, the lower the number of data analyzed. Small windows result in a higher definition of the time associated to frequency changes. However, smaller windows contain few data leading to a lower frequency definition. This corresponds to the uncertainty Heisenberg principle. Equation (3.63) tells us that, if the coefficient $d_{j,k}$ is large, then near time $t = 2^{-j} k f(t)$ oscillates with a wavelength proportional to 2^{-j} resulting in the *structure extraction* property of wavelets. Another wavelet property is the *localization*. If $f(t)$ shows a discontinuity at time t, only $\psi_{j,k}(t)$ overlapping that discontinuity will be affected, that is, only the coefficients of $d_{j,k}$ of that wavelet will be affected. Then, the wavelet transform provides not only the frequency information but also the time correspondent to the occurrence of that signal feature. This is another main difference with the FFT where discontinuities will affect all the FFT components. Let us consider again the expression of the CWT in equation (3.63). Here τ is the translation parameter and indicates the shift of the wavelet through the signal. The parameter s is a scale parameter. High values of s correspond to the general traits of the signal. Small values of s will describe the signal details. The DWT will be computed for every value of the discrete translation time τ for all the possible discrete scale value s.

Important is to observe that equation (3.60) is a convolution of the time sequence $x(t)$ for the wavelet $\psi^*[(t-\tau)/s]$, and this can be interpreted as a linear filtering operation. Essentially, algorithms computing the DWT of a given signal $x(t)$ produce two sets of coefficients the *scale coefficients* and the *wavelet coefficients*. The former are associated to the principal features of the signal and can be considered as the *coefficients of a low-pass* filter. The *wavelet coefficients* describe the signal details can be considered as the *coefficients of a high-pass* filter. Noise removal can then be performed selecting the desired number of *scale coefficients* which is the procedure applied in the *RxpsG* software. As done for the other filters, also for the wavelet filter, it is possible to compute the frequency response for a given degree of noise rejection. An example is shown in Figure 4.43 for a filter made of five wavelets

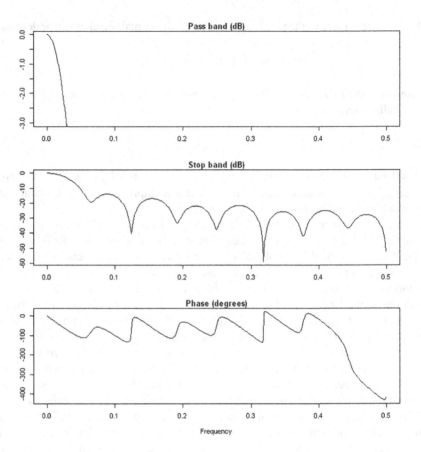

FIGURE 3.43   Frequency response of a wavelet filter composed by five base wavelets and with a rejection level = 5.

and noise rejection = 5. We will not enter in a more detailed description of the wavelet transforms and filter design which is out of the scope of the present book. We suggest the reader to refer to the wide literature on this topic [99–101], and for an introduction to wavelets and how they work the useful tutorial [102].

3.8.6.6.   FIR and IIR Filters

Desirable properties of a filter are the filter stability, the finite delay, and linear phase in the response. Finite impulse response (FIR) filters satisfy these requests. The simplest FIR filter is a moving average filter where the output is a simple average of a certain set of input data. More advanced FIR filters are characterized by a weighted average of a certain number of data using coefficients, the filter coefficients. The computation of appropriate filter coefficients allows to enormously improve the performances of the filter with respect to the simple moving average. The FIR filter design includes the selection of the number of coefficients which will correspond to the sharpness of the roll-off at the cutoff frequency and the attenuation in the stop-band. To better understand how an FIR filter works, let us do a digression. It can be demonstrated that a convolution of two arrays of data in the time domain corresponds to a multiplication in the frequency domain. Conversely, a convolution of a frequency spectrum in the frequency domain corresponds to a multiplication of the correspondent signal in the time domain. Filtering in the frequency domain corresponds to the suppression of some frequencies (as done, for example, in the case of FFT filters) and can be performed by simply multiplying all the frequencies of the pass-band by 1 and all the frequencies in the stop-band by 0 which can be seen as ideal transfer function of the filter. This multiplication in the frequency domain can be translated in the time domain by using an inverse FFT and corresponds to the convolution of the original signal by an impulse response of the filter. Then a generic signal can be filtered by convolving it by the impulse response of the filter. This is exactly what is done by applying a FIR filtering to a set of data and the impulse response of the FIR filter represents the FIR coefficients. The general form of an FIR filter can be described by the equation

$$y(n) = \Sigma_k h(k) x(n-k) \qquad\qquad 3.67$$

where $k = 0, \ldots, N-1$ and $h(k)$ are the filter coefficients. Equation (3.64) shows that the FIR filter has the form of a moving average filter as mentioned earlier. There are packages available to compute the FIR coefficients on the basis of the sharpness and noise rejection desired. Also R provides such an opportunity which was integrated in *RxpsG* smoothing routine. The user can verify the filter performances by filtering a step function. The filter frequency response is shown in Figure 3.44a for a 60 coefficients FIR filter with degree of noise rejection = 10. Please observe the flat response in the pass-band, the transition from the pass- to the stop-band, and the degree of noise rejection and the linear phase plot. However, to satisfy the degree of noise rejection and sharpness at the cut-off frequency, a rather high number of coefficients are required. This increases the undesired oscillations introduced by the filter in correspondence of signal discontinuities. Then a good balance of the number of coefficient and frequency response is required. In general for cleaning spectra form noise, a 20, 40 coefficients filter can be a good compromise.

An alternative to FIR filters are the infinite impulse response (IIR) filters. An IIR filter can be described by the equation

$$y(n) = \Sigma_k b_k x(n-k) + \Sigma_j a_k y(n-j) \qquad\qquad 3.68$$

where $k = 0, \ldots, M$ and $j = 1, \ldots, N$. The IIR filter resembles an autoregressive moving average filter where the first summation corresponds to the MA and the second summation to the AR part.

The frequency response of the filter is described by

$$H(z) = \Sigma_k b_k z^{-k} / (1 - \Sigma_j a_j z^{-j}) \qquad\qquad 3.69$$

The numerator of H(z) describes the filter zeros, while the roots of the denominator describe the filter poles which were not present in the FIR filters. The IIR filters better

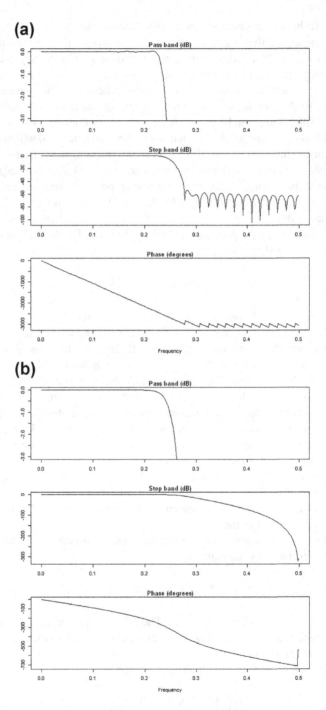

FIGURE 3.44   (a) Frequency response of an FIR filter with 60 coefficients and degree of noise rejection = 10. (b) Frequency response of an IIR with filter order = 8 and degree of noise rejection = 10.

describe the behavior of analog electronic filters composed of resistors, capacitors, and/or inductors. The presence of infinite impulse response introduces some signal distortion. Sharper cut-off filters can be obtained using IIR filters with a limited number of coefficients. However, the phase characteristic is not linear, and this can create problems to systems requiring phase linearity as it is shown in Figure 3.44b (bottom panel). In addition, IIR filters can display instability which can derive from its poles. Since the frequency response of FIR filters is described just by the numerator of H(z) in equation (3.66) (where b_k are the FIR coefficients), they do not have stability problems. In the case of IIR filters, one have to check if all the poles, the complex roots of the H(z) denominator in equation (3.66), lie inside the circle of radius = 1.

For both the filters, the number of coefficients defines the delay, namely the number of input data needed to compute the next filtered value. It corresponds to the number of coefficients used for the filter. In general, to obtain a filter response of all the elements of the signal, a padding is made. This operation corresponds to add a number of data equal to the number of filter coefficients N. At beginning of the x sequence, N elements = $x(1)$ are added while N elements = $x(n)$ are added at the end of the sequence. In the case of IIR filter, the number of padded data is increased to avoid filter oscillations introduced by the discontinuity at the end of the sequence $x(n)$. The main advantage of IIR filters with respect to the FIR filters is the easiest implementation, in order to meet the specifications in terms of pass-band, stop-band, ripple, and/or roll-off. The requirements can be met with a much lower order for IIR filter than for FIR filters.

3.8.7. CORE-LINE DERIVATIVE AND PEAK COMPONENTS

Sometime peak fitting is a complex procedure due to the difficulty of correctly recognize the contribution from different bonds when the correspondent fitting components are strongly overlapped. Several authors have proposed a variety of procedures for identifying the peaks underlying the measured spectrum. Some of these ideas can be tried to resolve the presence of components from a broad structure. The presence of a background can also interfere with the identification of the true structure. Differentiation is a commonly accepted method to identify the number of components of a spectral feature [60, 103–106]. Among the advantages of this method is the elimination of the contributions of backgrounds and slopes in the original spectrum. Also, quantitative analysis can be accomplished by comparing the intensities of second-derivative spectral bands. This method was seldom applied in IR spectra [107, 108], and it is common opinion that it does not provide accurate results in the case of overlapped spectral components as the curve-fitting method. However, the second derivative as preprocessing procedure can be helpful in the case of extremely broad peaks and when overlap is severe.

Generally differentiation of spectra leads to enhancement of the noise component with degradation of the signal-to-noise ratio as shown in Section 3.7.7. Nonetheless, it is possible to combine the advantages of second-derivative techniques in combination with curve-fitting methods [104]. As mentioned, the differentiation of the original data eliminates the effects of offset or slope. The first and second derivatives provide information to eliminate unnecessary components reducing the danger of data "overfitting". However, frequently first and second derivatives applied to the original signal are very noisy that the information about fitting components is completely obscured. A strong reduction of noise is needed applying high performing filters for noise rejection. The danger is the introduction of distortions which could affect the second derivative shape and consequent misinterpretation of the component number and position. Figure 3.45 shows an example of first derivative applied to a filtered spectrum from a SiO_x sample.

The filter was an IIR 6 coefficient filter with noise rejection 16. The first derivative indicates the presence of a fitting component with a positive-to-negative oscillation and the zero-crossing give indication of the correspondent BE position. However, in presence of overlapping, the shape of the derivative becomes more complex and must be correctly interpreted. Only the more pronounced oscillations must be considered disregarding those generated by the residual noise. In the example shown in

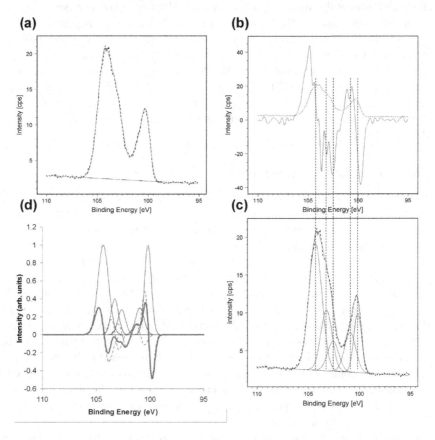

FIGURE 3.45   (a) Original (dashed line) and filtered (solid line) spectra from SiO_x sample. (b) The first derivative (solid) of the filtered SiO_x spectrum (dashed). (c) The fit (solid line) resulting placing fitting components (light solid) as indicated by the first derivative. (d) Synthesized first derivative (thick solid line) of a set of Gaussian components (thin solid) placed in the positions obtained from the Si 2p fit. Derivative of the single components are indicated with dashed lines.

Figure 3.45b, starting from the left side, the first derivative shows a main oscillation crossing the zero at the position of the first fit component. Then the next derivative oscillation is negative and split in two parts, indicating the presence of two fit components. Finally, follows a positive peak formed by two components, the last being placed at the crossing of the derivative with the zero line. Dashed lines indicate the positions of the fit components as derived from the first derivative, and used for fitting the Si 2p. The result of this procedure is shown in Figure 3.45c. Unfortunately, in presence of strong overlapping, the contributions from each of the fit components interfere and the first derivative cannot display separate positive-to-negative oscillations. This is illustrated in Figure 3.45d reporting the synthesis of the first derivative (red) of a sum of Gaussian components (blue). The components are located in the positions obtained from the Si 2p fit. In green are also shown the derivatives for each of the Gaussians. As it can be seen, when the derivatives are summed, in the middle region, the interference masks the positive-to-negative oscillations. The result substantially reproduces the main structures of the first derivative obtained by filtering and differentiating the original Si 2p data.

3.8.8. Loss-Features

The procedure to fit the loss features associated to a core-line is substantially equivalent to that illustrated for fitting the core-lines. Essentially, the *Analysis* option is used. Important to observe is that the spectrum is extended over a rather wide BE range. This enables the correct use of the Tougaard background which appears to be the more appropriate for the subtraction of the spurious contributions.

The lineshape of the loss features is rather complex and described in detail in [109] and [110]. The intensity of the loss features may be computed applying the Tougaard formalism to compute the cross section of the intrinsic and extrinsic surface and bulk plasmon losses [111]. Here we do not want to model the loss feature but trying to describe their overall shape to detect changes in the main spectral characteristics. For this reason, the loss features can be fitted using Voigt functions to describe the general trend of the structure intensity. Loss features may be fitted as shown, for example, in [112] to describe the gross changes of the system characteristics upon variation of the structural organization of atoms. For instance in the cited work, the analysis of the loss features was used to get information about the different hybridization of carbon atoms considering pure crystalline diamond and graphite as two prototypal references. There are cases in which the presence of specific loss features gives information about the chemical state of the correspondent element [113]. The cases of iron [114], cobalt [115], nickel [116], and copper (as already presented in Section 3.64) [117] are just a few examples of the use of loss features. As seen in these works, the loss features are fitted using Voigt functions as anticipated which can be selected using the *Analysis* option as indicated in Figure 3.46 for the Ni 2p core-line. In this example Ni was acquired on a SnONiO sample. Ni is then mainly in an oxidized form and only a very small component at ~852.7 corresponding to Ni^0 was found. The core-line is formed by the two spin-orbit components $2p_{3/2}$ $2p_{1/2}$. In the $2p_{3/2}$ spectrum, the main peak was assigned to NiO and $Ni(OH)_2$ (components C2 and C8), while the remaining components derive from loss features due to multiplet splitting (components C3 and C9, C5

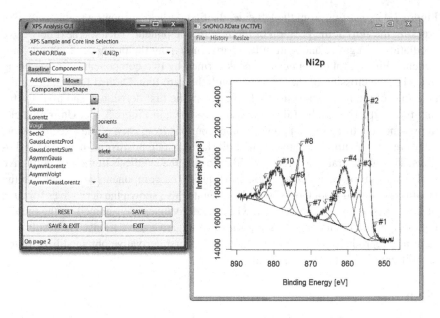

FIGURE 3.46   The *Analysis* GUI illustrating the selection of the Voigt linseshape for fitting the Ni 2p core-line acquired from a SnONiO sample. Ni is in an oxidized form; the various components at high BE represent the loss-features.

and C11) and surface/bulk plasmon excitations (the shake-up C4 component and the C6 with the correspondent spin orbit components C10 and C12). Presence of Sn will slightly affect the position of the fit components with respect to ref. [118].

The Ni 2p core-line was performed selecting a Shirley 3p background subtraction with a Distortion Parameter D = 0.5. The fit was performed using Voigt lineshapes linking the width of components 1 and 3 and components 2 and 4 to be the same. The result is shown in Figure 3.46 on the right. See Section 3.10 for additional information.

3.9.   AUGER SPECTRA

In 1922, Lise Meitner [119] discovered the electron emission process which could not be explained by a simple photoemission, and later independently the same process was discovered and explained in 1925 by Pierre Auger [120]. However, only with the work of J. J. Lander in 1953, the Auger emission excited by low-energy electrons, was recognized to be a useful technique for surface analysis [121].

The direct photoemission process leads to the creation of a hole in an inner core-level. Then there is a competition between two possible processes which originate from the decay of an outer electron to fill the inner hole: the *Auger Electron Emission* and the *X-ray Fluorescence*. These two processes correspond to the different ways the energy released by the electron relaxation is used by the atom. The excess of energy obtained when the inner hole is filled can emerge as an X-photon emitted by the atom. The second possibility is the excitation of an outer electron, the Auger electron, which will be emitted with a characteristic kinetic energy. The occurrence

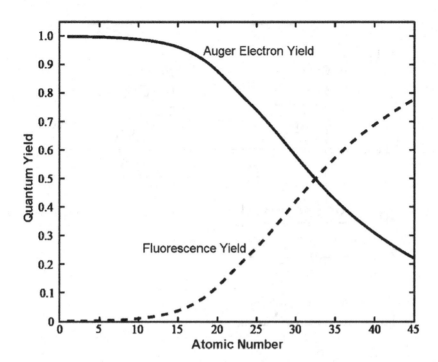

FIGURE 3.47 Quantum yields of Auger electrons and fluorescent X-rays per core-excited atom. For second-row elements, Auger yield represents >99% of relaxation events.

Source: Reprinted with permission from [122]

of one or the other of these two processes is governed by their probabilities which are shown in Figure 3.47.

As it can be seen, Auger emission is characterized by a higher probability for light atoms, while, for atomic number Z higher than 33, the X-ray fluorescence is favored.

3.9.1. THE AUGER PROCESS AND THE NOMENCLATURE

Let us enter in a more detailed description of the process. The Auger emission starts with the photoelectron emission already described in Section 1.2. Let us refer to the diagram in Figure 3.48. An incident photon of energy $h\nu$ is absorbed by an atom and a photoelectron is emitted with a kinetic energy equal to:

$$KE = h\nu - BE - \Phi$$

which is equation (1.3). Following the process of photoelectron emission, an ionized 1s core-level is created. The excess energy of this ion is released through a relaxation process in which a 2p electron from the upper L_2 shell fills the 1s hole, and then as observed earlier, either an X-ray photon $h\nu_x$ is emitted leading to X-ray fluorescence or a second electron is emitted through an Auger process. In both the cases holds the

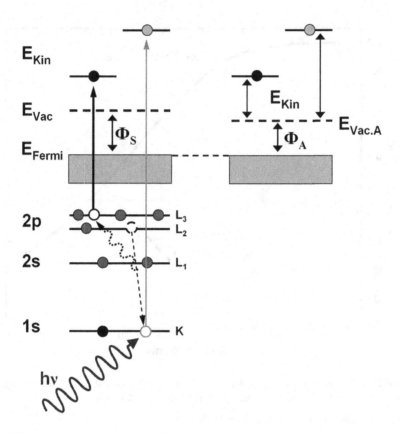

FIGURE 3.48 Schematic diagram of the process of Auger emission from a solid. An incident X-ray produces a hole in an inner energy level K. The energy released by the relaxation of an outer electron from the L_2 level, excites an electron of the level L_3 causing the emission of an Auger electron.

energy conservation law. In the case of X-ray fluorescence, the energy of the emitted photon simply equals the relaxation energy

$$h\nu_x = E_K - E_{L2} \qquad\qquad 3.70$$

where K indicates the 1s level in the spectroscopic notation. In the case of Auger electron emission the relaxation energy is directly absorbed by the atom leading to the emission of an Auger electron from an outer shell which in Figure 3.48 is indicated with L_3. The Auger transition will be conventionally indicated by using the spectroscopic annotation of the level involved in the process, KL_2L_3 in our case. The energy balance corresponding to this transition will be

$$KE_{Auger} = \left(E_K - E_{L2}\right) - E_{L3} \qquad\qquad 3.71$$

where E_{L3} corresponds to the BE of the level L_3.

As it can be observed, the energy of the X-rays used to ionize the core-level does not appear in equation (3.71), since Auger emission is generated by transitions between atomic levels (for instance the K and the L_2 levels in the above example). However, when wide spectra are acquired, Auger features are generally plotted on a binding energy scale. This makes the positions of Auger lines dependent on the X-ray source energy. For this reason, when only Auger spectra are acquired separately with a higher energy resolution, they are typically presented on a kinetic energy scale. Observe that the Auger process involves three electrons. Then, the Auger process can occur in elements having a number of electrons equal to or higher than 3 which corresponds to Li.

As seen earlier, the commonly accepted notation to indicate the Auger transition is the spectroscopic notation. This last is based on letters associated to the main quantum number followed by a subscript which discriminates levels with different orbital and spin quantum numbers. Table 3.4 summarizes the spectroscopic notation. In this formalism, the L-S coupling is assumed, l can take values $= 0, 1, 2, 3, \ldots s$ can assume values $S = -1/2, 1/2$ and the total angular momentum of the single electron j can take any value from $|l - s| < j < |l + s|$. Considering the possible values of the n, l, j quantum numbers the values listed in Table 3.4 are obtained.

Annotation of Table 3.4 is valid for atoms where the Russel-Saunders L-S coupling applies, that is, $Z < 20$. For atoms with higher atomic number, the j-j coupling must be applied. This brings to the following description.

The reader is invited to refer to specific textbooks [49, 123, 124] for a more detailed description of the spectroscopic nomenclature. If valence electrons are involved in the Auger emission process, the transitions are indicated using the letter identifying the initial ionized core-level, with the letter indicating the intermediate energy level and V for the valence electron. Then, for example, we will have KVV or $KL_{23}V$ transitions.

3.9.2. The Chemical Information

The chemical effects leading to shifts in the binding energy of photoelectrons affect also the Auger spectra. As described in the previous section, the kinetic energy of the Auger electron depends on the energy difference of the orbitals involved in the Auger emission process. By increasing the atomic number Z and the number of the electrons progressively filling the atomic orbitals, the binding energy of the atomic levels grows. However, the energy of the inner levels changes more slowly than that of the outer orbitals as shown in Figure 3.49a. Then the energy difference between inner and outer levels increases with increasing Z. As a consequence, also the energy released in the relaxation process raises with Z and consequently the energy relative to the KLL, LMM, MNN, etc. Auger transitions as illustrated in Figure 3.49b. An important outcome of this fact is that not only the core-lines but also the Auger spectra are element specific, allowing their identification. Figure 3.49b shows also that increasing Z increases the number of Auger possible transitions introducing complexity in the interpretation of the spectra. A partial example of this is shown in Figure 3.5 (Section 3.5). The peak in correspondence of the dashed line at ~1,100 eV describes the KLL transition of N 1s. The remaining structures are part to the LMM Auger transition of the elements 21–24 of the periodic Table. As it can be seen going from Sc to Cr, the Auger spectra become more structured. This derives by

TABLE 3.4
Spectroscopic Notation in the Russel-Saunders L-S Couplingn

n	l	j	X-ray suffix	X-ray level	Spectroscopic level
1	0	1/2	1	K	$1\,s_{1/2}$
2	0	1/2	1	L_1	$2\,s_{1/2}$
2	1	1/2	2	L_2	$2\,p_{1/2}$
2	1	3/2	3	L_3	$2\,p_{3/2}$
3	0	1/2	1	M_1	$3\,s_{1/2}$
3	1	1/2	2	M_2	$3\,p_{1/2}$
3	1	3/2	3	M_3	$3\,p_{3/2}$
3	2	3/2	4	M_4	$3\,d_{3/2}$
3	2	5/2	5	M_5	$3\,d_{5/2}$
...	...	...	...	...	...

TABLE 3.5
Spectroscopic Notation when Applies the j-j Coupling.

Transition	Configuration	L	S	Term
KL_1L_1	$2s^02p^6$	1/2	1	^1S
KL_2L_3	$2s^12p^5$	0	0	^1P
		1	1	^3P
$KL_{2,3}L_{2,3}$	$2s^12p^5$	0	0	^1S
		1†	1†	^3P†
		2	0	^1D
...	...	...	...	...

Note: † Forbidden because of the conservation of parity.

the increased number of orbitals satisfying the selection rules for a given transition. An additional problem is caused by the overlapping peaks in spectra collected from materials made up of several elements (cf. Figure 3.5, the N–KLL transition is overlapped to the LMM of the SC–Cr).

If a transition involves upper levels which are part of the valence band (VB), then the Auger spectrum carries information contained in the VB [127]. Since chemical bonding affects directly the valence levels, the Auger spectra will reflect differences in the element chemical state. As an example, Figure 3.50 shows the variation of the Auger spectra of Cu in its metallic neutral state, in its oxidized state, and when introduced in a glass by an ion exchange process.

The variation of the oxidation state and the influence of the environment directly influence the electronic structure of Cu and in particular the topmost energy levels. This is mirrored by the changes in the LMM Cu Auger spectra. This general property suggests the possibility of being able of recognizing different chemical states from

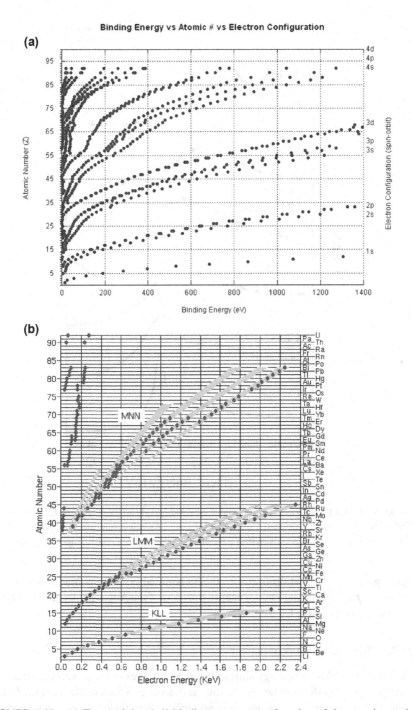

FIGURE 3.49   (a) Trend of the shell binding energy as a function of the atomic number. (b) Trend of the Auger transition kinetic energies ([a] Obtained from the website [125]; [b] Obtained from the website [126]).

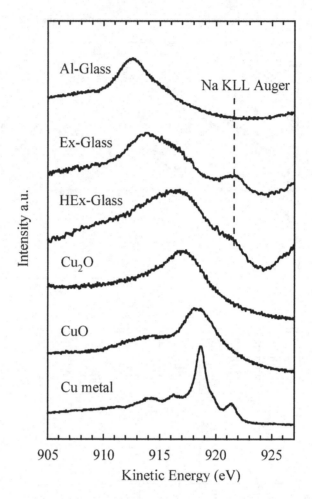

FIGURE 3.50 Cu LMM spectra of Al-glass, Cu exchanged glass (Ex-glass), and Ex-glass treated at 700°C (HEx-glass), Al enriched Cu exchanged glass (Al-glass), and the standard samples of CuO, Cu$_2$O, and Cu metal. A broken line indicates the position of Na KLL Auger main line.

Auger spectra. Identification of the element chemical bonding is made by referring to tabulated data listing the kinetic energy position of the Auger peaks [128, 129] and by comparison with spectra from literature.

3.9.3. THE AUGER PARAMETER

In addition to the kinetic energy shift induced by changes of the electronic structure, the Auger spectra offer another possibility to detect variations of the chemical environment of the atom. In particular, it is possible to compute the *Auger parameter* α defined as the energy difference between the kinetic energy (KE) of an Auger electron generated by a given relaxation to a core-level and the KE of a photoelectron from a core shell [130, 131]. Note that this energy difference is independent of the reference for the energy scale or of presence of charging effects. Also work function corrections are unnecessary

$$\alpha = E_k(c_1,c_2,c_3) - E_k(c_0) \qquad 3.72$$

where c_1, c_2, and c_3 are the core and outer energy levels involved in the Auger transition and c_0 is the core level from which the photoemission occurs. In the original definition of the *Auger parameter*, the core level of the Auger c_1 is different from c_0 involved in the photoemission but in fact generally they are the same. Following equation (3.72), the Auger parameter can assume negative values. It is then preferable to define a modified Auger parameter α' as the sum of the KE of the Auger electron and the excitation energy $h\nu$:

$$\alpha' = \alpha + h\nu = E_k(c_1,c_2,c_3) + E_b(c) \qquad 3.73$$

where $E_b(c)$ represents the binding energy of the initial core-hole c_0. From equation (3.73), it happens that variations of $\Delta\alpha'$ originate from changes in the core-hole binding energy $\Delta E_b(c)$ and in the kinetic energy of the Auger electron $\Delta E_k(c_1,c_2,c_3)$. By definition, the parameter α' is independent of the energy of the X-photon excitation. The first term can be written as [132, 133]:

$$\Delta E_b(c_0) = \Delta\varepsilon(c_0) - R^{ea}(c_0) \qquad 3.74$$

the variation of the core binding energy is the variation of the core-level energy induced by the change in the electronic charge, and $R^{ea}(c_0)$ represents the extra-atomic relaxation energy. As for changes of the kinetic energy of an Auger electron, this is given by the difference between the initial and final state. For core-type Auger lines, considering that the Auger kinetic energy is described by $E_k(c_1,c_2,c_3) = E_b(c_0) - E_b(c_1,c_2) - E_b(c_2)$

$$E_k(c_1,c_2,c_3) = \varepsilon(c_0) - R^{ea}(c_0) - 2\varepsilon(c_2,c_3) + R^{ea}(c_2) + R^{ea}(c_3) - R^s - F(x) \quad 3.75$$

where R^s is the intra-atomic relaxation energy for the c_2 and c_3 levels, while $F(x)$ is the electron–electron interaction due to the double vacancy in the final state. Considering that changes in the environment substantially do not influence R^s and $F(x)$, equation (3.75) leads to

$$\Delta E_k(c_1,c_2,c_3) = \Delta\varepsilon(c_0) - R^{ea}(c_0) - 2\Delta\varepsilon(c_2,c_3) + R^{ea}(c_2,c_3) \qquad 3.76$$

The chemical shifts in different shells are experimentally closely the same, that is, $\Delta\varepsilon(c_0) \approx \Delta\varepsilon(c_2,c_3)$. In addition, it can be demonstrated that for a one-hole and two-hole state, the relaxation or polarization energy for $R^{ea}(c_2,c_3) = 4R^{ea}(c_0)$ [131] and then

$$\Delta E_k(c_1,c_2,c_3) = -\Delta\varepsilon(c_2,c_3) + 3R^{ea}(c_0) \qquad 3.77$$

It follows that

$$\Delta\alpha' = \Delta E_k\left(c_1, c_2, c_3\right) + \Delta E_b\left(c_0\right) = \Delta E_k\left(c_1, c_2, c_3\right) + \Delta E_b\left(c_0\right) = 2R^{ea}\left(c_0\right) \quad 3.78$$

Equation (3.78) tells us that the differences in photoelectron and Auger chemical shifts originate from the difference in final state, extra atomic relaxation energies between chemical states. The extra-atomic relaxation is connected with the charge distribution in the final state. For the given system, the most important change is due to the charge transfer from the neighboring atoms to the photoionized atom.

Equation (3.73) indicates that the Auger parameter values are the intercepts of the linear relationship $\Delta E_k\left(c_1, c_2, c_3\right)$ versus $E_b\left(c\right)$ with slope -1. Different chemical states for a given element will result in different α' values. With respect to the chemical shift displayed by the core-lines, the parameter α' is more sensitive, because the double vacancies in the upper level result in a twofold relaxation energy R^{ea} (equation 3.78). In Figure 3.51, it is shown an example of α' Wagner plot for barium compounds (see ref. [134] for a complete list). Different symbols are used in Figure 3.51 to identify groups of barium compounds as oxides, halids, sulfides, non-metallic anions, and Ba metal. In Figure 3.51 are plotted also straight lines with slope -1. According to equation (3.78), higher Auger parameter values are related to higher values of R^{ea} (extra-atomic contribution). Equation (3.73) is susceptible to assume a different form when similar initial state effects, that is, the potential V_M generated by the neighboring atoms (Madelung potential) and of the ground state valence charge q of the core-ionized atom. In this case

$$E_k = \left[const + 2\left(V_M + kq\right)\right] - 3E_b \quad 3.79$$

Equation (3.79) shows that compounds with similar initial state effects will appear in the Wagner plot on straight line with slope -1. More information may be found in [133].

3.9.4. THE SECOND DERIVATIVE OF THE AUGER SPECTRA AND BAND STRUCTURE RECONSTRUCTION

Auger spectra are commonly represented in two different ways on the kinetic energy scale. The first is the direct representation of the spectral intensity as acquired by the detector. This method includes not only the contribution of the Auger electrons but also that of the background due to secondary electron scattering. The second method applies the first derivative to the acquired signal to suppress the slow varying background and display just the contribution deriving from the Auger processes. However, differentiation has a wider application not limited to emphasize spectral features suppressing the background. As seen in Section 3.7.8 for core-lines, differentiation is a method applied to recognize the presence of spectral components to correctly perform peak fitting. Similarly, the second derivative was also proposed for the analysis of Auger spectra to describe their structures [135]. In addition, the sensitivity of the second derivative to detect spectral features was utilized to describe

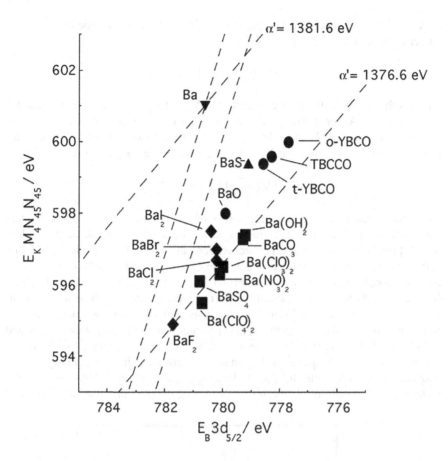

FIGURE 3.51 Wagner plot for barium compounds. Data plotted versus core-level $3d_{5/2}$ binding energies, Auger $M_4N_{45}N_{45}$ kinetic energies, and Auger parameter for materials containing Ba. Auger parameters for BaF2 and elemental Ba are reported on lines with slope equal to 1. For the same compounds, lines with slope equal to 3 are also reported. Each compound in the plot may have its own couple of lines with slope equal to 1 and to 3: (▼) Ba metal; (♦) halides; (■) non-metallic anions; (●) oxides; (▲) sulfide.

Source: Reprinted with permission from [134].

the Density of State (DOS). An example is the analysis of the KVV spectrum derived from carbon described in ref. [136]. We will take this work as an example which can be extended to elements other than carbon. Observe that changing element not only changes the Auger spectrum, but different effects at higher order may enter at play (many body effects). Then, each case needs to be carefully analyzed. Authors of [136] demonstrated that the negative of the second derivative of the C-KVV spectrum displays oscillations corresponding to the DOS components. This possibility comes out by remembering equation (3.71) when, as in the present example, the upper levels are valence levels. Then for a carbon atom, the kinetic energy of an Auger electron may be described as

$$E_K = E(1s) - E(V_1) = E(V_2) \qquad\qquad 3.80$$

$E(1s)$ is the energy of the C 1s core level and V_1 and V_2 represent the valence states involved in the Auger transition. A simplification can be made considering simply $V_1 = V_2 = V$, and then $E(V_1) + E(V_2) = 2E(V)$ instead two different valence states [137, 138]. Then from equation (3.80), we can describe the energy of a valence state as

$$E(V) = \left[E(1s) - E_K\right] / 2 \qquad\qquad 3.81$$

Using this relation and experimental values of the 1s core-level of carbon, authors were able to reproduce its band structure in good agreement with theoretical calculations [136, 139, 140]. The Auger second derivative was further analyzed by other authors [141] for understanding to which extent a simple self-fold of DOS can provide a good description of the C KVV spectra and the effect of spectral distortions by many body effects. At this aim, a careful design of preprocessing which includes deconvolution of loss features and signal filtering was adopted to remove noise. Extrinsic losses were deconvoluted from the Auger spectrum before differentiation as indicated in [142–144]. The negative of the second derivative is then compared with the one obtained from the VB spectrum. Important is to consider that originating from the convolution of n VB levels, the C KVV Auger spectrum is described by n (n+1)/2 possible states. The results of the spectral analysis done in [145] are shown in Figure 3.52

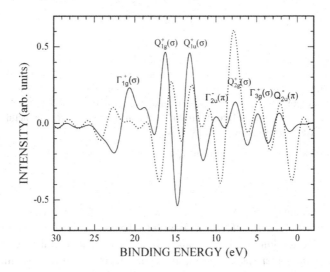

FIGURE 3.52   Second derivative (with opposite sign) of the VB photoemission (continuous line) and CVV Auger emission (dotted line) spectra from Highly Oriented Pyrolitic Graphite (HOPG). The energy of the features is marked on the figure.

Source: Reproduced with permission from [145]

The reproduction of the band structure from VB and C KVV shows substantially the same number of features with a very good overlap in energy position in the spectral portion toward the Fermi Level while some differences are observed at ~ 11 and 15.5 eV and at the DOS bottom. In the first case, the differences can arise by the different spectral sensitivity (cross section) to s- and p-electrons [145]. Likely the differences at high BE derive from the breakdown of the single-particle approximation used to describe the C KVV spectrum. In particular, presence of repulsive interaction between the two holes in the final state leads to a reduced kinetic energy for the Auger electron. The extension of the DOS obtained from the second derivatives allows an estimation of the hole-hole interaction of ~ 2 eV in a very good agreement with theoretical calculations [146].

3.9.5. ANALYZING THE AUGER SPECTRA

The number of possible Auger spectra, also for light elements, is composed by a certain number of transitions, rendering difficult the interpretation of the spectra. Fortunately, most of these lines are too weak to be detected. An estimation of the expected transition intensity can be obtained by multiplying the number of the electrons in the three orbitals involved in the Auger emission. A qualitative description of the Auger spectrum may be obtained following the model proposed in [147]. Concerning the energies of Auger electrons, it is possible to obtain a good approximation using the empirical rules

$$E_{c1} = \left[E_{c1}(Z) + E_{c1}(Z+1) \right] / 2$$

$$E_{c2} = \left[E_{c2}(Z) + E_{c2}(Z+1) \right] / 2$$

$$E_k = E_{c0}(Z) - \left[E_{c1}(Z) + E_{c1}(Z+1) + E_{c2}(Z) + E_{c2}(Z+1) \right] / 2 \qquad 3.82$$

where c_0, c_1, and c_2 identify the core level and the upper levels involved in the Auger transitions (cfr. equations 3.71, where levels L_2 and L_3 correspond to c_1 and c_2, respectively). When the energy difference between states c_2 and c_3 is small, equation (3.82) can be approximated by [148]

$$E_k = E_{c0}(Z) - E_{c1,2} \qquad 3.83$$

where $E_{c1,2} = E_{c1}(Z) + E_{c2}(Z+1)$ identify the external orbitals. In this approximation, the outer electrons of the atom Z containing an inner core-hole see a potential generated by the inner electrons approximately equal to that in atom Z + 1 with the extra electron removed. An example of calculated data for chromium is reported in Table 3.6 and is in good agreement with experimental results. It is possible to give an estimation also of the expected relative Auger intensities. At this aim, authors in [147] used the product of the number of electrons in the three levels involved in the Auger transition. These values were then normalized to 100 assigned to the maximum among the transition intensities for a given element and called them "normalized multiplicities".

TABLE 3.6
Calculated Auger Spectra for Chromium Levels (*)

Orbital	Population	Z Energy (eV)	Z+1 Energy (eV)
K	2	5989.00	6539.00
L I	2	695.00	769.00
L II	2	584.00	652.00
L III	4	575.00	641.00
M I	2	74.00	84.00
M 23	6	43.00	49.00
M 45	5	2.00	4.00
N I	1	3.00	3.50

TABLE 3.7
List of Auger Lines between 10.0 and 3,000.0 eV for Initial Vacancies up to 6,000 eV and Having *Normalized Multiplicity* Greater than 0 for Cr Element

Element	Vacancy level	Interaction levels		Energy (eV)	Normalized multiplicity
Cr	M I	M 23	N I	24.75	3
Cr	M I	M 23	M 45	25.00	40
Cr	L I	L II	M 23	31.00	16
Cr	M 23	N I	N I	36.50	4
Cr	M 23	M 45	N I	36.75	20
Cr	M 23	M 45	M 45	37.00	100
Cr	L I	L III	N 23	41.00	32
Cr	M I	N I	N I	67.50	1
Cr	M I	M 45	N I	67.75	6
Cr	M I	M 45	M 45	68.00	33
Cr	L I	L II	N I	73.75	2
Cr	L I	L II	M 45	74.00	13
Cr	L I	L III	N I	83.75	5
Cr	L I	L III	M 45	84.00	26

Table 3.7 lists the normalized multiplicities relative to the Auger transitions of chromium.

The values of *normalized multiplicity* give a crude but helpful means to identify the most intense Auger transitions in experimental spectra. However, it has to be considered that the relative peak intensities depend on ionization cross section, width, and fluorescence probability. Then estimations may deviate from experimental data [149] rendering sometimes difficult to identify elements in composites by using only tabulated data. However, the general tracts of the Auger spectra are reproduced and this can help the identification of the elements. Also, and more important, handbooks [128, 129, 150] are available for standard reference elements enabling a reliable identification of chemical species.

Element	Vacancy level	Interaction levels		Energy (eV)	Normalized multiplicity
Cr	L III	M I	M I	417.00	10
Cr	L II	M I	M I	426.00	5
Cr	L III	M I	M 23	450.00	32
Cr	L II	M I	M 23	459.00	16
Cr	L III	M 23	M 23	483.00	96
Cr	L II	M 23	M 23	492.00	48
Cr	L III	M I	N I	492.75	5
Cr	L III	M I	M 45	493.00	26
Cr	L II	M I	N I	501.75	2
Cr	L II	M I	M 45	502.00	13
Cr	L III	M 23	N I	525.75	16
Cr	L III	M 23	M 45	526.00	80
Cr	L II	M 23	N I	534.75	8
Cr	L II	M 23	M 45	535.00	40
Cr	L I	M I	M I	537.00	5
Cr	L III	M I	N I	568.50	2
Cr	L III	M 45	N I	568.75	13
Cr	L III	M 45	M 45	569.00	66
Cr	L I	M I	M 23	570.00	16
Cr	L II	M I	N I	577.50	1
Cr	L II	M 45	N I	577.75	6
Cr	L II	M 45	M 45	578.00	33
Cr	L I	M 23	M 23	603.00	48
Cr	L I	M I	N I	612.75	2
Cr	L I	M I	M 45	613.00	13
Cr	L I	M 23	N I	645.75	8
Cr	L I	M 23	M 45	646.00	40
Cr	L I	M I	N I	688.50	1
Cr	L I	M 45	N I	688.75	6
Cr	L I	M 45	M 45	689.00	33

Source: Obtained with permission from [147]

3.9.6. BACKGROUND SUBTRACTION AND AUGER FITTING

The procedure for background subtracting and fitting the Auger spectra is the same as that for the core-lines and it is based on the *Analysis* option already described in Section 3.7. Depending on the material, and following the indications already given for the core-lines, background subtraction is first performed. Figure 3.53 shows a Shirley background subtraction for a carbon KVV Auger spectrum from an HOPG sample. Then, as normally done, fitting components are added to describe the spectrum. In the present example, are used seven Voigt components. Observe that the

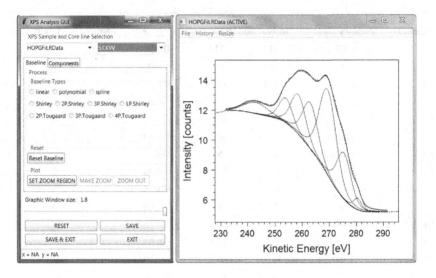

FIGURE 3.53 Auger KVV spectrum of a Highly Oriented Pyrolitic Graphite (HOPG). A Shirley background subtraction and Voigt components are used for spectral fitting.

number and the position of the components are not casual. They derive from the band structure of HOPG. Figure 3.53 represents the best fit of the C KVV spectrum obtained by placing the fit components in the position indicated in ref. [145]. Those positions were maintained fixed during the fitting procedure. Also the width of the components was not allowed to vary freely. Initially, all the component FWHM were forced to assume the same value. Then, looking at the band structure of HOPG and considering many body effects on the tail at low kinetic energies, the components were left to vary in a reasonable range around their initial value which was ~2.5 eV.

3.9.7. First- and Second-Derivative of the Auger Spectra

Modern instruments operate in pulse counting mode yielding the intensity I (counts per second) as a function of the kinetic energy E. However, Auger spectra are frequently visualized using their first derivative. There are historical reasons for using the derivative spectra. In old analog systems, the derivative spectrum was directly obtained from lock-in amplifiers. However, still the first derivative is used for the advantages provided. Differentiation damps slow variation of the signal resulting in the elimination of the background. Also, in differentiated spectra, the peak-to-peak height is used as a measure of the intensity of the non-differentiated Auger peaks. Finally, differentiation emphasizes signal variations and then the effects of the different chemical bonds. Spectral differentiation is an easy operation since in digitalized signals it expression is simply the difference between adjacent data.

$$y'_n = y_n - y_{n-1}$$

3.84

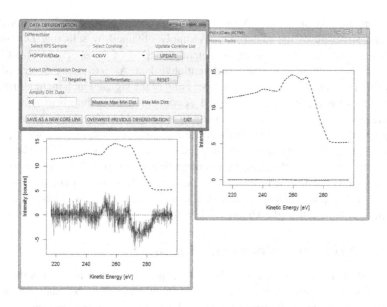

FIGURE 3.54   The Differentiate GUI on the top left. On the right, original data and the first derivative of C KVV with its original amplitude are shown. The bottom-left plot displays the same derivative multiplied by a factor 50.

Derivatives of higher order are obtained just re-applying the difference 3.83

$$y'' = y'_n - y'_{n-1}$$ 3.85

In *RxpsG*, these operations are performed using the option *Differentiation* illustrated in Figure 3.54.

When using the GUI *Differentiation* after selection of the *XPS-Sample* and of the core-line, the user must indicate the degree of differentiation, 1 in the example. Then just pressing the button *Differentiate*, the derivative of the spectrum is obtained. Because the spectral derivative is the difference between adjacent data, its amplitude is small as seen in Figure 3.54 in the top-right plot. To visualize the derivative together with the original data, the user can amplify the amplitude of the derivative inputting an *amplification factor* which, in the present example is 50 as indicated in the GUI at the top-left of Figure 3.54. Despite the high SNR of the original data, the derivative is rather noisy as it can be seen and better results can be obtained filtering-out the noise as described in Section 3.7.7. This is needed to obtain reasonable values of the distance between derivative maximum and minimum which is representative of the width of the Auger features. This parameter is used, for example, to discriminate different hybrids of carbon as described in [151] and [152]. This measure is readily done by pressing the button *Measure Max Min Dist.* (bottom-right of the GUI) which activates a manual estimation done by mouse-clicking the max-min positions directly on the first derivative spectrum.

The second derivative of the Auger spectrum is computed to get information about the band structure of the material. It is simply obtained by selecting *Differentiation degree* equal to 2. However, in general the SNR of the spectrum is not sufficiently high to obtain reasonable differentiated data. As previously described (see Sections 3.77 and 3.84), a careful preprocessing of the signal which includes filtering and loss feature deconvolution is needed. A convenient procedure is to select a filter and clean the spectrum. Save the results in a *Smoothing Test* core-line and try to compute the second derivative. If the noise rejection is not sufficient, apply a more severe filtering and overwrite the results in the *Smoothing Test* core-line. Then check if the data are sufficiently clean by re-computing the second derivative on the new filtered data. If needed, as in the case of the second derivative, it could be useful to deal with the negative of the differentiated data. This can be readily done by marking the option *Negative* near the *Differentiate* button.

3.9.8. DECONVOLUTION OF THE LOSS FEATURES

In general, there are not ideal signals. Data acquired by an instrument is the superposition of several effects including secondary effects inside the analyzed system, perturbations of the environment affecting the system, the response function of the instrument, and noise deriving from the electronics. All these phenomena affect the original signal, thus modifying the acquired data. The convolution is a mathematical operation on two functions $f(x)$ and $g(x)$ that produce a third function $C(x) = f(x) \otimes g(x)$, mirroring how the shape of f is modified by g. The mathematical expression of the convolution $f(x) \otimes g(x)$ is

$$C(x) = f(x) \otimes g(x) = \int_{-\infty}^{\infty} f(x)\, g(x-t)\, dt \qquad 3.86$$

The convolution is used to describe the mixed effect of many different sources on the experimental data. The discrete form of the convolution is expressed as a summation

$$C(n) = \sum_{m=1}^{M} f(m)\, g(n-m) \qquad 3.87$$

where m runs on the number of elements of f. The convolution can be performed using two different methods. The first is the direct computation of the products for each of the n elements of $g(n)$. The second method is to use the property of the FFT transform. It can be demonstrated that, the in the frequency domain, the convolution product $C(x) = f(x) \otimes g(x)$ is just the product of the transformed functions:

$$FFT(C(x)) = FFT(f(x)) \cdot FFT(g(x)) \qquad 3.88$$

Then taking the inverse of the FFT transform, one obtains

$$C(x) = iFFT(FFT(f(x)) \cdot FFT(g(x))) \qquad 3.89$$

The *XPS Convolution* offers both the two possibilities. The procedure starts with the selection of the XPS Sample, of the *core-line 1*, and of the *core-line 2* to convolve. Then the user must select the desired convolution method and press the relative button to make the calculation.

Observe that we are assuming that the two spectra are acquired in the same conditions, namely the same energy step and represented on the same binding or kinetic energy scale. Before convolving/deconvolving two spectra, it is then necessary to ensure that these requirements are respected. When required, the user can utilize the options *XPS Interpolate Decimate* (see Section 6.4) in the *Analyze* main menu to modify the spectrum energy step and, if needed, use the *Switch KE/BE* function in the *Plot* main menu, to change the energy scale. As usual, the GUI allows the selection of the XPS Sample and the correspondent core-lines see Figure 3.55A. Just as an example, the spectra to be convolved can be a C KVV Auger spectrum and an extended C 1s spectrum with its loss features. Both these spectra are baseline subtracted using the *Analysis* GUI using a simple Shirley and a Tougaard baseline respectively. Once the two spectra are selected the *Convolution* can be readily computed using the FFT or the sum of products: equivalent results must be obtained. The button *Save & Exit* adds the spectrum to the XPS Sample.

To explain how the *XPS Deconvolution* procedure works, it is necessary to give a piece of information regarding what done in the past. Fine analysis of the Auger features requires the deconvolution of the loss components which affect the spectrum. This work was done long time ago using the inverse of the FFT (iFFT) [143] or using alternative methods [142, 144] or the Van Cittert algorithm [153, 154]. The application of the inverse of the FFT for deconvolving the loss features leads to spectra strongly affected by noise. A pre-filtering of the original spectra allows the iFFT to produce a more reasonable result although filtering can also remove information. The Van Cittert iterative approximation [153, 154] is an alternative but its application is not straightforward. A common severe problem inherent to these works derives from the convolution. If n is the number of elements of A and m is the number of elements of B by definition, the convolution $C = A \otimes B$ is formed by $n + m$ elements. Then one cannot deconvolve the loss features from the Auger transitions and obtain a spectrum with the same number of points as the original one. This problem is solved in the *XPS Deconvolution* procedure.

The user can select the *iFFT* or the *Van Cittert* algorithm to perform the operation. Although the original C KVV and C 1s spectra have a rather good SNR (see Figure 3.55b), just noise will be obtained as a result of the deconvolution (Figure 3.55c). Let us consider the *Van Cittert* procedure. This algorithm minimizes the difference obtained by subtracting from the original Auger spectrum its estimate obtained by convolving the same Auger and the C 1s spectra. The process is iterative: at convergence the difference is minimized and the convolving spectrum is obtained as a solution of the process (for more details, see [154, 155]).

Upon selection of the XPS Sample and of the core-lines, a message will appear asking if the FWHM of the core-line 2 is much smaller than the FWHM of the core-line 1 (see Figure 3.55). This information is needed, because the algorithm behaves differently if the widths of the two spectra are similar or markedly different

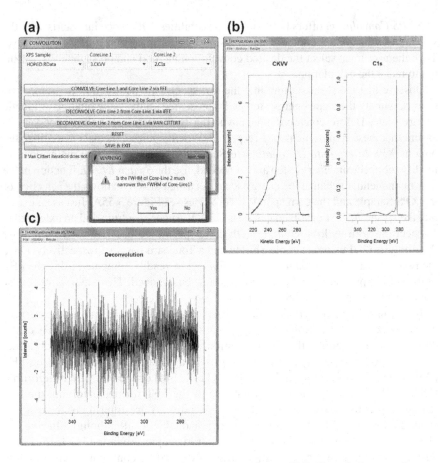

FIGURE 3.55   (a) The *XPS Convolution Deconvolution* GUI. Convolution may be performed by using both the direct FFT algorithm (much faster) and the summation of nested products (slow for long arrays of data). Deconvolution is computed applying the inverse of the FFT or the *Van Cittert* algorithm. (b) Different widths of the CKVV and C 1s spectra allow application of the van Citter method. (c) Result of the deconvolution of the C 1s from the CKVV spectrum applying the iFFT algorithm.

as in the present example. This diversity derives from the following property of the deconvolution

$$\delta(x) \otimes f(x) = f(x) \qquad\qquad 3.90$$

where $\delta(x)$ represents the Dirac function $\delta(x) = 1$ if $x = x_0$, $\delta(x) = 0$ otherwise. If the core-line 2 is very narrow, it behaves similar to a delta function. This can be easily verified by checking the FWHM of C KVV $\otimes$ C 1s which appears almost the same as that of the original C KVV spectrum as illustrated in Figure 3.56. Then the remaining parts can be disregarded. This solves the problem of the different lengths

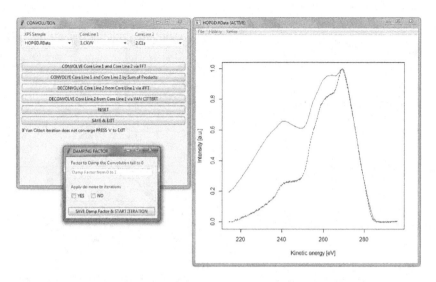

FIGURE 3.56 The selected portion of the C KVV ⊗ C 1s with length L (solid line) is compared to the original Auger spectrum (dashed line). As it can be seen, cutting the tail at low kinetic energies leads to a convolution different from zero at the edge.

of the convolution and of the Auger spectrum. However, as reported by Figure 3.56, trimming the tail of the convolution introduces a non-zero edge at low kinetic energy.

To harmonize this edge with the Auger spectrum, a *Damping Factor D* is required. By increasing *D* the convolution tail is smoothly reduced to zero as illustrated in Figure 3.57a, b. Now the iterative *Van Cittert* algorithm can be applied and the result is shown at each cycle of the iteration.

Comparing Figures 3.57c and 3.57d, it is possible to observe an increase of the spectral noise. This is a drawback of the *Van Cittert* algorithm deriving from the iterative procedure. A partial limitation of this problem is obtained selecting the *Denoise* option in the *Damping Factor* window as illustrated in Figure 3.57b.

3.10. VALENCE BANDS

Formation of chemical bonds involves the topmost orbitals of atoms. Photoelectron emission from those orbitals should directly reflect the formation of the bonds and the changes of atom's chemistry. The photoelectron spectra obtained from outer molecular orbitals give rise to *Valence Bands* (VB). Because molecular orbitals are directly involved in the formation of chemical bonds, the VBs should be very sensitive to any change of the atom's bonding when analyzing different molecules and this sensitivity includes also structural changes from amorphous to crystalline structures. More in general, we can say that VBs describe the top electronic configuration of the materials and for this reason they are analyzed to reconstruct the band structures and the density of states which can be compared to those from theoretical models (see, e.g., [139, 140, 156]) to shed light on the physical, chemical, optical properties of materials.

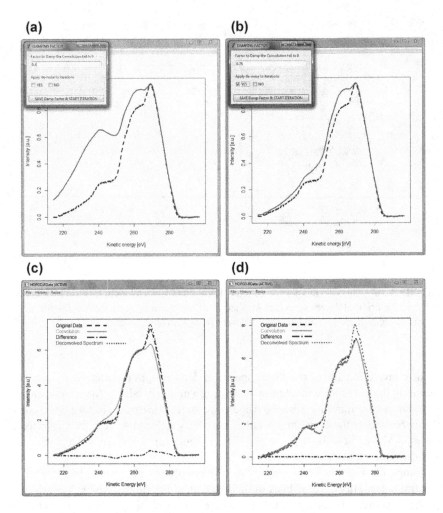

FIGURE 3.57   (a) Variation of the convolution edge for *Damping Factor* value = 0.4 or (b) 0.75. In (c), the result of the *Van Cittert* procedure applied to the harmonized convolution. In (d), the result of the *Van Cittert* iteration: at beginning (c) and at convergence (d), the convolution (solid line) overlaps the original data (dashed line), the difference (dashed-dotted line) is almost null leading to the final deconvolved spectrum (dots).

3.10.1. The Chemical Information

As pointed out, valence bands reflect the atoms bonding, thus mirroring the material surface chemistry. This in principle is true although valence spectra are formed by bands generally broader than core-lines. Generally, the interpretation of valence band requires high quality spectra and very careful computation using elaborate quantum mechanical models which should reproduce any type of experimental data. To show the different sensitivities in probing different portions of the electronic structure, let us consider as an example the core-lines and VBs of

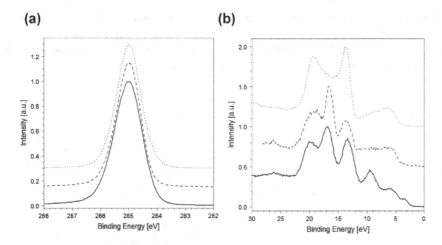

FIGURE 3.58 (a) C 1s core-line from polystyrene (solid line), polypropylene (dashed line), and polyethylene (dotted line) and (b) correspondent valence bands.

TABLE 3.7
Monomer Structure of Polyethylene (PE), Polypropylene (PP), and Polystyrene (PS)

PE   PP   PS

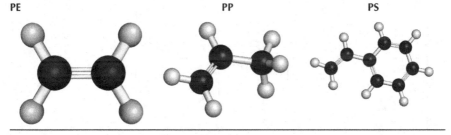

Figure 3.58. Here are reported the spectra from polyethylene (PE), polypropylene (PP), and polystyrene (PS).

As it can be observed, the three polymers display similar C 1s core-lines, formed just by a peak at the same BE with no particular additional features. This result can be understood considering that in all the cases carbon is bonded only to hydrogen and then no substantial differences can be seen on the basis of the chemical shift. However, the chemical structure of the three polymers is different as reported in Table 3.7. Correspondently, Figure 3.58b shows that the VBs of these three polymers are different.

Indeed, VB mirrors the electronic structure of the outer orbitals, and different monomers as those of PE, PP, and PS lead to different electron arrangements, thus explaining the changes in the VB photoemission spectra. This agrees with theoretical computations [157] where different electronic structures were found for these three polymers as shown in Figure 3.59. This is a clear indication that the monomer chemical structure and their more or less ordered packing in an amorphous or semicrystalline bulk material have a deep influence on the outer electronic configuration of a polymer and then on the spectral features and on the band gap.

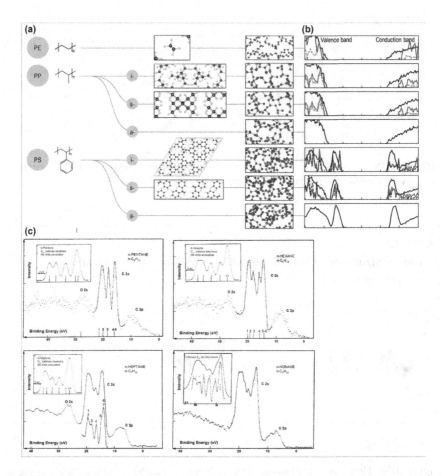

FIGURE 3.59 Relationship between the physicochemical and the electronic structures of six model polymers. (a) Structures of the monomers and the crystalline and the amorphous phases of the polymers considered, with the different tacticities (i.e., isotactic [i-], syndiotactic [s-], and atactic [a-]). We note that the tacticity information for certain crystalline polymers is unknown and only the local backbone structure of amorphous phases is shown. The C, H, and O atoms are denoted by gray, white, and red spheres, respectively. (b) The total electronic density of states (TDOS) and the projected density of states (PDOS) corresponding to CH2–CH2, CH2–CH(CH3), and benzene groups in the crystalline phases. The energy levels are with respect to the average C-1s core level of the crystalline PE. (c) Change of the DOS by increasing the nu,ber of carbon and hydrogen atoms from pentane to nonane molecules. The VB approaches that of polyethylene.

Source: Adapted with permission from [157]

This enhanced sensitivity of the VB to detect various types of isomerism (structural-, linkage- and stereo-) as well as to tacticity and geometrical conformation was already pointed out in studies at beginning of 1980 [158]. It is instructive to follow how the VB of PE can be explained on the basis of periodicity and broadening of the energy levels of simple alkane molecules [158]. Successive molecules with an increasing number of atoms were studied as progressive steps for the formation of bulk PE. Authors of [158–160] show that initially the VB is formed by well-distinct C 2s and C 2p regions in the CH_4 molecule. Increasing the number of atoms as in n-C_9-H_{20} molecule results in increasing of the density of states, leading to a spectrum very similar to that of PE as shown in Figure 3.59c. Similar trends are found for other polymers as illustrated in [158].

Interpretation of the VB is made with the knowledge of the chemical structure of the material analyzed. In the case of pure elements, the VB will be formed by bands deriving from the topmost element orbitals. Atoms packed in a crystalline or amorphous structure lead to a multiplicity of levels falling in a limited energy range thus forming the bands. Depending on the principal quantum number, the HOMO, the highest occupied molecular orbital, level can be a p orbital or, in the case of transition elements, d or f orbitals. However, presence of orbital hybridization or mixing leads to a less distinct separation of the bands characterized by different orbital angular momentum L. When composites are analyzed, the VB becomes even more complex since the contribution of all atoms fall in the same energy range. Nevertheless, the analysis of the VB can be performed recognizing the contribution of the different single elements by comparison with appropriate reference materials, considering the formation of the molecular orbital of the separate elements and by comparing spectra with theoretical modeling.

3.10.2. VALENCE BAND FROM DIFFERENT MATERIALS

In this section are described some examples of VBs acquired on different materials to illustrate the different kind of information obtained analyzing these spectra. The VB can be obtained from same molecules in both gas and solid phases. Figure 3.60 shows the comparison between free molecules and the condensed phase to highlight the effect of the intermolecular interactions and crystal-field effect developing in bulk matrixes [159]. The example in Figure 3.60 regards the VB from gaseous and solid nonane molecules.

The two spectra show similar structures apart from an energy shift to higher BE typical of gaseous phases with respect to the solid one. Essentially the two spectra display a more prominent band correspondent to the C 2s and a less intense C 2p band at lower binding energy. The bigger difference between the two spectra arises from the broadening of the band components in the solid phase which makes C 2s and C 2p structures less distinguishable with respect to those of the gas phase. However, the broadening is limited. In fact, the VBs maintain a similar shape increasing the length of the hydrocarbon molecules till to the formation of polyethylene which can be considered approximately as one-dimensionally bonded carbon system (cfr spectra of nonane in Figure 3.60 with that of polyethylene in Figure 3.61). The comparison can be further extended to other carbon systems as graphite and diamond

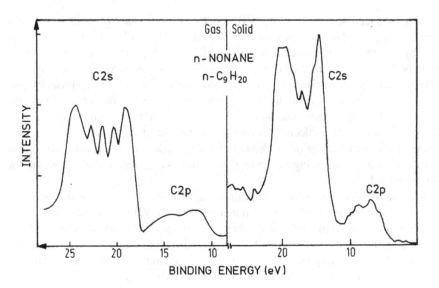

FIGURE 3.60   Valence-electron spectra of n-nonane, recorded in the gas and solid phases.
—*Source:* Reproduced with permission from [159]

where, apart from the bond with hydrogen, differences in carbon hybridization and consequent crystalline structure come at play. The spectra of these prototypal materials are shown in Figure 3.61. As identified by the different line patterns, the VBs are composed by bands mainly derived from 2s and 2p orbitals in agreement with theoretical calculations [140, 161, 162]. In the center, the main peak originates by a superposition of the s and p contribution. Finally, in the case of graphite, in the region adjacent to the Fermi level falls the contribution deriving from π electrons [139, 140]. The intensity of this feature is rather low due to the low cross-section of p, π electrons toward X radiation. The X-ray photoemitted VB can be complemented using UV photons leading to a VB where the p, π regions are enhanced thanks to the higher cross section. The s and p character of the VB can be investigated using polarized X-photon from synchrotron radiation [163]. With p-polarization, the electric field is very near to the axis of the spectrometer and maximizes the intensity from s-orbitals. By contrast, the photoemission from these orbitals is suppressed using s-polarization where the field is almost perpendicular to the analyzer axis. This allows to better visualize the s and the p regions of the diamond VB spectrum and in particular the region proximal to the Fermi level [163].

The VB spectra can mirror complex relaxation effects occurring in emission from outer orbitals. An example is the VB of transition metal dihalides [164] as NiO and $NiCl_2$. These materials were analyzed using He I and He II radiation to better identify the contribution deriving from p and d orbitals thanks to the different p, d cross sections for these photons. If we apply the Koopmans' theorem and neglect relaxation effects, the ionization energy (binding energy) of the d orbitals is always larger than that of the p orbitals. Experimental results are in contrast to this picture, because $d_{\pi g}$

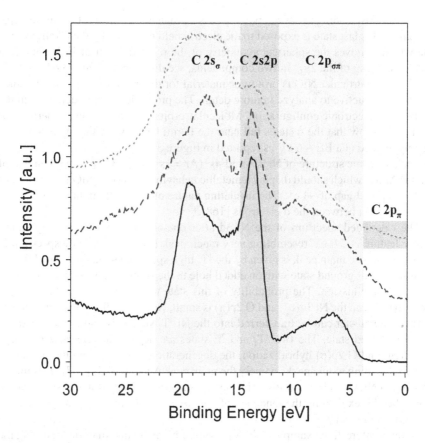

FIGURE 3.61 Comparison of VB spectra deriving from pure sp³ systems as polyethylene (solid line) and diamond (dashed line) and pure sp² system as HOPG (dotted line).

and $d_{\sigma g}$ are located at smaller binding energies than the p orbitals, and only the $d_{\delta g}$ orbital is placed below the p orbitals. To explain these results, one has to conclude that relaxation effects are important [165]. In particular, the $d_{\pi g}$, $d_{\sigma g}$ have the same symmetry of their counterparts in the chlorine p orbitals. Therefore, a charge transfer from the p-to-d orbitals can occur making relaxation possible. On the other side, because there are no p orbitals of $d_{\delta g}$ character, charge transfer is hindered and no relaxation effects can occur in this case. If the previous systems are characterized by covalent bonds, alkali salts can be considered as prototypes of ionic bond compounds [166]. In these systems, the electron charge is strongly localized around the more electronegative atoms. With the loss/acquisition of an electron charge, both the elements assume an electronic configuration similar to that of the noble gases. For example, in Li⁺F⁻ the electronic configuration is Li⁺ ($1s^2$) and the F⁻ ($1s^2 2s^2 2p^6$). The photoemission process can take place only in the negative ion because extraction of a second electron from the alkali atom would require an energy equal to the second ionization potential. After the photoemission the system will be Li⁺F⁰ the ionicity of the bond is lost leading to a weak bond. F^0 has a p state with a spin-orbit splitting $2p_{1/2}$ and $2p_{3/2}$. This last state is exposed to the electric field produced by the alkali positive ion which removes the spherical symmetry of the p electrons in of F^0 leading to a further splitting of the $2p_{3/2}$ in two components, see diagram in Figure 3.62.

Ni and in particular NiO is prototype material for the transition metal compounds which is instructive to analyze in more detail. The properties of NiO derive from the Ni peculiar electronic configuration. Molecular-orbital calculation and experimental data [167] show that the d states fall near the Fermi level, while the O 2p states are mainly located at a BE ~ 6 eV as reported in Figure 3.63.

The electronic structure of Ni is Ar $3d^8 4s^2$ (Ar = $1s^2\ 2s^2 2p^6\ 3s^2 3p^6$) and then that of NiO is Ar $3d^8$ which should display a metallic behavior. However, NiO is an insulator with an optical gap of ~4 eV. The insulating nature of NiO is thought to result from the correlation between the d electrons [168].

The calculated spectrum of the Ni 3d after one-electron removal clearly shows a new feature at 3.0 eV resembling very much peak B of the bulk NiO spectrum in Figure 3.63. The main peak is given by the $^4T_1$ high spin state which is formed from a high spin Ni d^8 ground state with an added hole in the spin-down t_{2g} orbital produced by the photoemission. The probability of this state to disperse is low, because the overlap between the Ni $3d(t_{2g})$ and O $2p(\pi)$ is small. Now an e_g electron from a neigh-boring NiO cluster can be transferred into the Ni $^4T_1$ site. This leaves that neighbor in a ^2E ionized state. The two $^4T_1$ and ^2E states are degenerate. Because the large Ni $3d(eg)$ and O $2p(\sigma)$ hybridization, the degeneration of the $^4T_1$ and the ^2E state is removed leading to two peaks namely the main peak A and peak B [167]. This model shows that NiO is a charge-transfer insulator with a spin-compensated first ionization state, thereby explaining the tendency of Ni^{3+} oxides to form low-spin systems and its antiferromagnetic behavior.

Another interesting example is the VB from semiconducting materials. Among the systems gaining interest for their electronic and optical properties, ZnO has gained

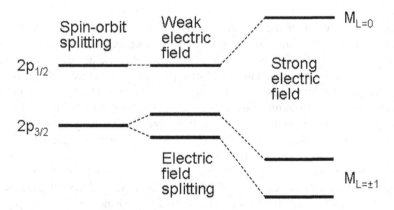

FIGURE 3.62 Diagram showing the spin-orbit splitting of the F 2p orbital and the effect of the weak or strong electric-field generated by the positive Li^+ ion resulting in a further split-ting of the $2p_{3/2}$ level in two $M_{L=0}$, $M_{\pm 1}$ states.

the attention of the scientific community for the possible applications in blue and ultra-violet optical devices, sensors, solar cells, catalysis, and new nanotechnology based devices [169]. Depending on the preparation method and the system conformation (macro- or nanosize, shape and aspect ratio, etc.) [170], ZnO assumes an energy gap which is in the range of 3.1–3.4 eV [171] and these values may change upon doping. Depending on the kind of dopant, the band gap (BG) of ZnO may increase or decrease. As an example Mn or Cd doping results in a decrease of the BG [172] while an opposite effect is obtained using Mg [173]. In [171], authors studied pure and Cu-doped ZnO. They found that annealing their system at increasing temperatures (from 400 °C to 1200 °C) results in a decrease of the BG from ~3.33 eV to 3.19 eV. Same annealing process was applied to $Zn_{0.99}Cu_{0.01}$ system. Presence of Cu reduces the BG values which now range from 3.3 eV to 2.98 eV. Besides the optical analysis of the materials, authors performed XPS to check both the chemistry and the electronic structure. At higher annealing temperature corresponds an increase of the hydroxyl groups in both pure and doped ZnO and then a different kind of bonding

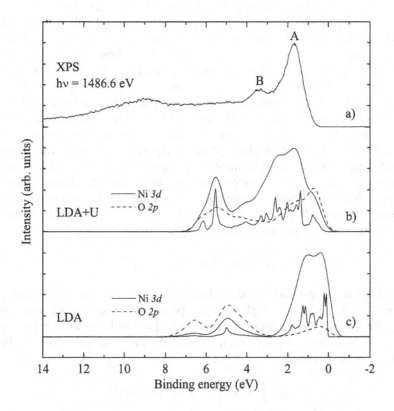

FIGURE 3.63   (a) Valence band (XPS) (1486.6 eV) spectrum of an in situ cleaved NiO single crystal together with the results of (b) LDA + U and (c) LDA calculations (black curves), with and without the broadening to account for the experimental energy resolution.

Source: Reproduced with permission from ref. [167]

of oxygen to Zn and Cu elements. The inspection of the VB of pure and doped ZnO reveals three features the more intense deriving from Zn 3d at ~10 eV while the features at lower BE in the range 8–0 eV derive from O 2p and hybridization of Zn 3d–O 2p orbitals. The main effects on the VB are as follows: (i) reduction from macro- to nano-scale of the system size leads to an increase in the band gap from 3.19 eV to 3.32 eV, which is accompanied by a blue shift of the VB edge of 0.41 eV. (ii) The same effect is found in Cu-doped ZnO where micron-sized particulate display a 2.98 eV BG, while in the nano-sized particulate, the BG is 3.28 eV. Also in this case, the reduction of the system size leads to a blue-shift of the VB edge of 0.32 eV. (iii) Cu doping causes a reduction of the BG both in micro-systems and in nano-systems. (iv) In micro-sized and in nano-sized ZnO particulate, the doping leads to a red-shift of the VB edge of 0.12 and 0.03 eV, respectively.

As a last example, we will describe the SnO_2 system. Pure SnO_2 is a direct BG semiconductor, the excitation of electrons from the VB to the CB occurs from the Γ symmetry point of both the VB and CB. The doping of SnO_2 with Sb leads to a band gap increase due to the VB downward shift of the VB and an opposite upward shift of the CB. The VB of pure and doped SnO_2 is reported in Figure 3.64.

The SnO VB displays three main peaks deriving from the hybridization of Sn 5p–O 2p orbitals (the two structures nearer to the VB edge) and the Sn 5s–O 2p orbital at ~ 11 eV. Doping with Sb causes a downward shift of these three peaks to higher BE. In undoped SnO_2, a less intense feature is present at low BEs. This seems to derive from substoichiometric Sn reduction from Sn^{4+} to Sn^{2+}, causing a mixing of the Sn 5p–5s orbitals [174]. A less pronounced tail is present also in the Sb-doped SnO_2 which originates from the segregation of Sb^{3+} ions at surface where they replace Sn^{2+} cations. This hypothesis is sustained by the intensity of the Sb core-level spectra which is higher than the expected on the basis of the Sb doping concentration. The evaluation of the VB edge is performed by a linear extrapolation (Figure 3.64) which locates the VB onset at ~ 3.60 eV and 3.95 eV with respect to the Fermi level for undoped and Sb-doped SnO_2, respectively. Authors of [174] were also able to estimate the width of the occupied part of the conduction band. The loss features are induced by interaction with conduction electrons, and the energy of the plasmon features

$$\omega^2_p = Ne^2 / \left[m * \varepsilon_0 \left(\varepsilon_\infty + 1 \right) \right] \qquad 3.91$$

where N is the concentration of the conduction electrons with effective mass m*, e is the electron charge, ε_∞ is the high-frequency permittivity, and ε_0 the vacuum permittivity. The carrier concentration may be estimated on the basis of the doping level. Because the effective mass depends linearly on the carrier concentration, the Fermi level E_F can be described as

$$E_f = \int_0^\infty h^2 dn / \left[4 \, (3\pi^2 n)^{1/3} \left(m_0^* + cn \right) \right] \qquad 3.92$$

where c is a constant. Equation 3.92 allows for the evaluation of the populated CB width. Considering that the energy shift of ~0.45 eV induced by Sb (cfr Figure 3.64) is much lower than the value expected for the calculated width of the occupied CB, a shrinkage is hypothesized. This effect is attributed to the attractive Coulombic potential

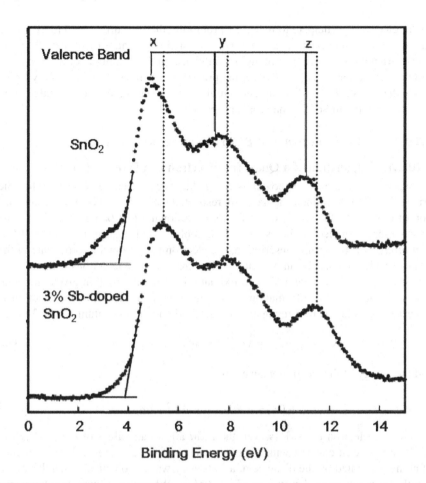

FIGURE 3.64 Valence-band XPS of undoped SnO_2 and 3% Sb-doped SnO_2. Linear extrapolations of the valence-band edges are shown: these allow estimates to be made of the valence-band onset energies. Vertical lines highlight shifts in the valence band peaks x, y, and z with doping. The peak positions were determined by fitting the valence-band profile to a series of Voigt peaks. See text for details.

Source: Reproduced with permission from [174]

associated with the higher nuclear charge of Sb atoms as compared with Sn leading to an attractive dopant-electron interaction. A second effect derives from the conduction electrons screening the repulsive interaction between conduction and valence electrons thus stabilizing the CB states. The band-gap renormalization for the carrier concentration of $\sim 4 \times 10^{20}$ cm^{-3} is estimated to be in the range 0.05–0.2 eV in quite good agreement with experimental results.

These examples show how, differently from core-lines which are generally used to shed light of the chemical bonding of atoms, the analysis of VBs sheds light on the electronic configuration of atoms with high detail. This is fundamental when outer electrons are involved in any kind of processes, from those involved in catalysis where electrons participate to reactions promoting their occurrence, to those where electrons are used in electronics, photovoltaics and in photonics. In this case the electronic structure, the effect of doping, its influence on the amplitude of the band-gap and the consequent optical response, the presence of defects inducing energy levels inside the gap, or the band alignment in multi-junction cells are fundamental to draw a satisfactory picture of the material's properties.

3.10.3.   BAND STRUCTURE AND BAND STRUCTURE RECONSTRUCTION

3.10.3.1. Propagation of a Quasi-Free Electron in a Periodic Potential

Crystalline solids are characterized by energy bands whose energy changes when the orientation of the specimen changes with respect to the analyzer axis. For the description of this effect, the reader can refer to the following text books [175–179]. The propagation of an electron in a solid is commonly made by assuming the *Free Electron Model (FEM)*. In this description, the electron is described to propagate in the conduction band in a constant potential as if there were no interactions with the environment. This is of course a big approximation but the description given, although crude, provides useful information about the system. In the *FEM*, the electron is described as a plain wave propagating in the solid with a momentum $\mathbf{p} = \mathbf{h} / \lambda = \hbar \mathbf{k}$

$$\psi(k, r) = e^{ik\,r} \qquad\qquad 3.93$$

and the energy of the electron is given by

$$E(k) = \hbar^2 k^2 / 2m = \hbar^2 k^2 / 2m_e \qquad\qquad 3.94$$

with m_e the electron mass. However, the *FEM* allows all values of k and therefore of E which are in contrast with the experiments. If we consider a linear sequence of atoms separated by the distance $\mathbf{a}$, a stationary wave is obtained when the wavelength $\lambda = n\mathbf{a}$ or equivalently when $k = 2\pi/n a$. When this occurs, the plain wave assumes the same periodicity of the Bravais lattice. The set of k satisfying the relation $k = 2\pi/na$ is known as *Reciprocal Lattice*. The *Reciprocal Lattice* is defined by base vectors which are defined orthogonal to the crystalline plains of the conventional crystalline lattice. If $\mathbf{a}$, $\mathbf{b}$, $\mathbf{c}$ are the primitive vectors of the Bravais crystalline lattice, the primitive vectors of the *Reciprocal Lattice* are defined by

$$A = 2\pi / a(\perp b, c) \qquad B = 2\pi / b(\perp a, c) \qquad C = 2\pi / c(\perp a, b) \qquad 3.95$$

where $\perp \mathbf{b}$, $\mathbf{c}$ means normal to the plain defined by vectors $\mathbf{b}$ and $\mathbf{c}$, and similar for the other. It follows that the product of a generic vector of the crystalline lattice $\rho = n_1 \mathbf{a} + n_2 \mathbf{b} + n_3 \mathbf{c}$ for a vector of the reciprocal lattice $G = m_1 \mathbf{A} + m_2 \mathbf{B} + m_3 \mathbf{C}$ is a multiple of 2π and then $\exp(i\,G \cdot \rho) = 1$. The unit cell defined by the vectors $\mathbf{A}$, $\mathbf{B}$, $\mathbf{C}$ is also called *First Brillouin Zone*. The first *Brillouin Zone* is the equivalent of the Wigner–Seitz cell in the reciprocal space. In both the cases, the unit cell contains only a single point of the lattice. Conventionally the center of the *Brillouin Zone* is the center of the unit cell so that they extend from $-\pi / a$ to π / a, $-\pi / b$ to π / b,

$-\pi/c$ to π/c. Successive *Brillouin Zones* are simply obtained by taking multiple of the primitive reciprocal vectors. For example, the second *Brillouin Zone* is defined by $2*\mathbf{A}$, $2*\mathbf{B}$, $2*\mathbf{C}$ corresponding to vectors G varying in the range $[-\pi/\mathbf{a}, -2\pi/\mathbf{a}]$, $[\pi/\mathbf{a}, 2\pi/\mathbf{a}]$, and similarly for $\mathbf{b}$ and $\mathbf{c}$. The properties of the vectors of the reciprocal lattice are particularly convenient because vectors of *Reciprocal Lattice* can directly be associated to the momentum of the particle propagating through the crystalline lattice as we will see. As for the Bravais lattice where do exist periodicity relations among the coordinates x, y, z of a generic point

$$\mathbf{r}' = \mathbf{r} + n_1 u_x + n_2 u_y + n_3 u_z$$

where u_x, u_y, u_z are the vectors defining the lattice unitary cell, a similar relation holds for the *Reciprocal Lattice*

$$k' = k + 2\pi/g_x + 2\pi/g_y + 2\pi/g_z \qquad 3.96$$

where $g_i (i = x, y, z)$ represents the *Brillouin Zone* in the tree spatial dimensions. Let us return to the description of an electron emitted in a crystalline matrix; its energy can be described by

$$E_k = h\nu - E_B - f \gg \hbar\, k^2/2m \qquad 3.97$$

Equation (3.97) shows that the energy of the photoemitted electron depends on the wave vector k which is quantized following the *Brillouin Zone*. If the momentum of the electron $\mathbf{p} = \hbar k = \hbar \mathbf{g}$, it is easy to see that it satisfies the Bragg diffraction condition

$$2a\, \sin\theta = n\lambda \qquad 3.98$$

where $\mathbf{a}$ indicates the separation between adjacent lattice planes while λ corresponds to the wavelength of the photon or electron wave function. Looking at Figure 3.65, Bragg diffraction occurs when

$$k' - k = \Delta k = 2k\, \sin\theta$$

Since $k = 2\mathring{A}/\lambda$ follows that the Bragg condition is satisfied when

$$\Delta k = 2n\pi/a = g$$

or

$$k' = k + g \qquad 3.99$$

Equation (3.99) represents a momentum conservation law being g the amount of momentum exchanged with the crystalline lattice. Equation (3.99) tells us that

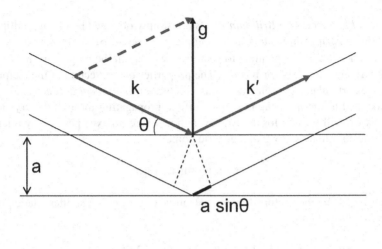

FIGURE 3.65   Bragg reflection occurs for a multiple of the wavelength equal to *2a sinθ* or equivalently for $k' = k + g$.

crystalline plains act as mirrors reflecting the wave functions associated to electrons which then are described by stationary waves as those of a harmonic oscillator. Consider that the Bragg constraint on the wavelength must correspond to a constraint on the momentum remembering that $\mathbf{p} = \hbar k = 2\pi / \lambda$. Follows that the spectra observed along different crystalline directions may change. Then acquisition of spectra varying the specimen orientation, with respect to the X-excitation and the acquisition (analyzer) directions, allow reconstructing the band structure of the material.

The crystalline structure can be thought as an ordered structure of positive ions surrounded by a negative charge. The *FEM* can be improved adding the effect of a periodic potential generated by the electron charges and positive ions. Now the electron wave functions are plane waves modulated by the crystalline periodic potential (Bloch theorem)

$$\psi(k, r) = e^{ik \cdot r} u(k, r) \qquad 3.100$$

These wave functions are called *Bloch functions* and the term *u(k, r)* describes the periodic potential. Because the crystalline field follows the symmetries and structure of the crystalline lattice, the wave functions are modulated by this potential with the same symmetries and periodicities. It is easy to demonstrate that the additional term *u(k, r)* must satisfy some properties

$$u(k, r + a) = u(k, r) \qquad 3.101$$

where **a** represents the periodicity of the lattice. As for the *Free Electron Model*, the description of the behavior of an electron in a periodic potential allows for connecting

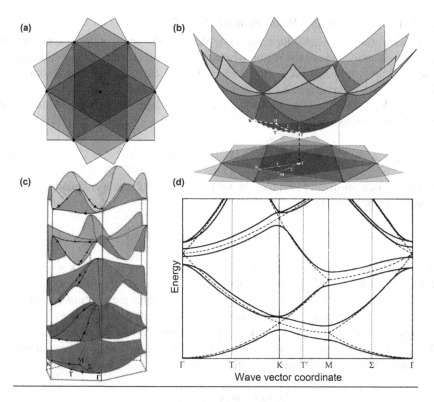

FIGURE 3.66 (a) Brillouin zones for a hexagonal crystalline lattice. Different colors are used to indicate the sequence of the five Brillouin zones. (b) An energy paraboloid is obtained from the Brillouin zones remembering that $E = \hbar \, k^2/2m$. (c) Reduced Brillouin zone. The black trace connects the main symmetry points which can be observed by orienting the specimen along the symmetry axis of the crystalline structure. (d) Band structure obtained moving along the indicated symmetry points on the five Brillouin zones.

the final state k_f values to those of the initial states thus probing the electronic structure of the material at different orientations. Let us consider, as an example, a material with hexagonal crystalline structure (which is the structure of crystalline graphite, crystalline cadmium, cobalt, zinc, etc.). For a two-dimensional lattice, the *Brillouin Zones* are obtained by connecting a starting point, we define as the origin, of the real crystalline lattice with all the neighbors. Then for each connecting line, we consider the planes passing for the midpoint and normal to that line. It can be demonstrated that those plains satisfy the conservation of the momentum $k = k' + g$. The intersection of the plains defines the unit cell in the reciprocal space. For a 2D-hexagonal lattice, the result of this operation is depicted in Figure 3.66a. Experimental evidence of this sequence of *Brillouin Zones* can be found in [180]. The different colors indicate the first five different *Brillouin Zone*. The energy expressed by equation (3.97) is a function of k^2. As a first approximation, Figure 3.66b shows also the energy paraboloid correspondent to the two-dimensional planar sequence of the five *Brillouin Zones*.

3.10.3.2.   Electron in a Weak Periodic Potential

The description given holds for a quasi-free electron. The edges of the *Brillouin Zones* touch each other and are the region where, for appropriate value of the momentum, the electron is reflected back. However, in the real case, the electron is immersed in a potential V_0 generated by the distribution of ions and electrons of the crystal. Because V_0 has the same periodicity of the crystalline structure, it can be described by a Fourier series

$$V_0(r) = \Sigma_{n \neq 0} V_m(k_m) \exp(i \, k_m \cdot r) \qquad 3.102$$

It may be demonstrated [177] that the solution of the Schrödinger equation in presence of the potential V_0 is

$$E = \hbar^2 k^2 / 2m \pm V(k = g_i) \qquad 3.103$$

In correspondence of all k-vectors satisfying the Bragg conditions $k = g_i = \pm \pi / a$, the energy splits into two separated levels correspondent to the Fourier component of the potential for that k vector. Referring to Figure 3.66c, the edges of the *Brillouin Zones* are separated and, as a consequence, also the correspondent energy bands. An energy gap corresponding to $2V(k = g_i)$ is formed. Periodicity of the potential V leads to a discrete translational invariance of the Bloch functions. If T_a is a translation operator which applies a shift of a lattice spacing **a**

$$T_a \psi(x) = \psi(x + a) = e^{ika} \psi(x) \qquad 3.104$$

Equation (3.104) tells us that the eigenvalue of T_a is a phase e^{ika}. Observing that $e^{ika} = e^{i(k + 2\pi/a)a}$ states identified by **k** or $k + 2\pi / a$ are the same. This enables us to restrict our description to the region $k = [-\pi / a, \pi / a]$ which identifies the *First Brilluoin Zone* (Figure 3.66c for the hexagonal lattice). However, if the potential *V* is not too high (a too high potential will confine the electron in a potential well corresponding to the lattice unit cell), the first term of the Fourier expansion prevails on the other and the expression of the wave function remains similar to the original Bloch function 3.99. The result of the Bloch theorem 3.99 is unexpected: the presence of a periodic potential does not result in a complete localization of the electrons in some regions of the crystalline lattice. Differently, the electrons are still described by plane waves modulated by a periodic function $u_k(x)$, but their wavenumber is restricted to the first *Brillouin Zone*. Accepting this picture, we can shift all the *Brillouin Zones* with wave vector extending behind the $[-\pi / a, \pi / a]$ range into the first one by applying a shift of *ng* (n = 1, 2, 3 . . .). This means that we will observe pure optical transitions as vertical transitions connecting energy branches of the first *Brillouin Zone*. If exchange of energy with the lattice is present, then the original wave vector will be shifted in **k′** + **g** to respect the momentum conservation.

It is important to observe that, in presence of interactions, the momentum of the electron described by the Bloch function k_{Bloch} is not constant. The simple relation $E = \hbar p = \hbar k_{free}$ (k_{free} is the wave vector associated to the free electron) is no more valid. However, we may still consider that $\hbar k_{Bloch}$ describes an average momentum of the electron composed by two terms: the momentum of the free electron $\hbar k_{free}$ and a second term deriving from the interaction of the electron with the lattice. The expectation values of the "Bloch momentum" are given by

$$p = m / \hbar \nabla_k E \qquad\qquad 3.105$$

which is the expression of the *group velocity*. The entity $\hbar k$ does not represent the simple momentum of the electron as it was for the free electron so that k is no longer equal to the expectation value of the momentum. Then k is no more directly connected to the electron kinetic energy. $\hbar k$ is called the *crystal momentum* of an electron and accounts for the interaction of the electron with the crystalline lattice and the exchange of energy due to scattering processes which can be elastic or can involve generation or extinction of phonons. In this latter case holds the momentum conservation law. If an electron k in a collision interacts with a phonon of wave vector q, equation (3.99) of the momentum conservation becomes

$$k + q = k' + g \qquad\qquad 3.106$$

In other words, an electron absorbs a phonon q and scatters from its initial state k to the final state k' by exchanging a momentum g with the lattice.

3.10.3.3. Energy Bands and Photoemission Spectra

Let us consider the radiation-matter interaction leading to the photoemission. For the description of the process, we here are assuming the *Three Step Model*. The first step corresponds to the excitation of the electron above the vacuum level; the second step corresponds to the propagation of the photoelectron through the bulk to the surface; finally, the third step consists in overcoming the surface potential barrier and escaping into the vacuum. The interaction between a radiation and electrons of a solid is described by a Hamiltonian of the form

$$H_{int} = e / mc\, A \cdot p \qquad\qquad 3.107$$

where A represents the vector potential orthogonal to the propagation direction of photons. We will neglect the momentum carried by the photon as we will justify in the following, and consider excitation with conservation of the momentum. What happens during the photoemission process is described in Figure 3.67a. Emission from the surface occurs for excited electrons whose component of the kinetic energy normal to the surface is sufficient to overcome the surface potential barrier. All the other electrons are reflected back into the bulk. Inside the crystal, the electron travels in a potential of depth $E_v - E_i$ where E_v is the vacuum energy level while E_i is the energy of an initial state $|i\rangle$. The emission from the surface occurs only if the energy satisfies the condition

$$\hbar^2 k^2_{\perp} / 2m \geq E_B + \Phi \qquad\qquad 3.108$$

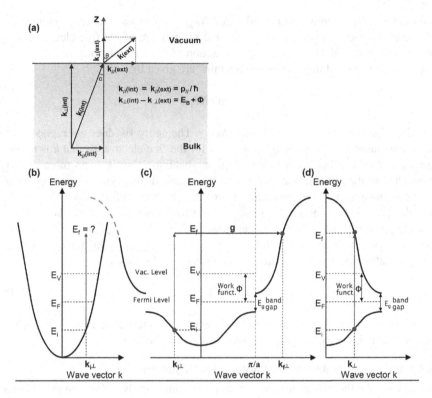

FIGURE 3.67 (a) Refraction of the electron momentum during bulk-to-vacuum emission. Only the parallel component $k_{//}$ is conserved in absence of interactions with phonons. (b) Optical transition in the case of free electron. The energy is represented by just a parabola where it is not possible to have transitions without variations of k. (c) Absorption of a photon (with negligible k_p) results in a transition from an initial state $|i\rangle$, k_i to a final state $|f\rangle$, k_f mediated by a phonon g. The transition is represented in an *extended Brillouin Zone*. (d) Because of the quantization of phonon energies, it is possible to describe the electron transition in a *Brillouin reduced scheme*. The transition from an initial state $|i\rangle$ to the final state $|f\rangle$ occurs with a defined *lattice wave vector k*.

$E_B = E_F - E_i, \Phi = E_v - E_F$. As indicated in Figure 3.67a, $k^\perp$ represents the component of the electron momentum normal to the sample surface. As for $k_{//}$, the parallel component of the momentum holds the relation

$$k_{//}(\text{ext}) = k_{//}(\text{int}) \qquad\qquad 3.109$$

Relation 3.108 tells that, crossing the solid–vacuum interface, the parallel component of the wave vector is conserved as shown in Figure 3.67a. In a photon-electron interaction, the conservation of the momentum should include the momentum associated to the photon so that $k_{f4} = k_{i4} + g + k_p$ where k_p represents the wave vector of the photon (observe that normal components of the wave vectors are used). However, the momentum of the photon p is very small with respect to that of the electron. For photons in the visible, λ is of the order of 10^3 Å, while the bond lengths are limited to few angstroms. For example, for X-photon energy of 1486.6 eV, the wavelength is ~8.3 Å resulting in a k_p ~ 0.7 Å^{-1} while typical reduced wave vectors are ~ 2 Å^{-1}. Increasing the wavelength to He II UV photons, k_p ~ 0.02 Å^{-1} and for visible photons k_p ~ 0.001 Å^{-1}. As a consequence in optical transitions the momentum carried by the photon is neglected. In a free electron system optical transitions are not possible because the lack of appropriate final states as sketched in Figure 3.67b. In presence of a periodic potential, the electron is scattered and the transition from an initial state $|i\rangle$ identified by k_i to the final state $|f\rangle$ identified by k_f is mediated by a wave vector $\mathbf{g}$. The crystalline lattice acts as a source and as a sink of momentum. In other words, the lattice acts as a buffer allowing conservation of energy and momentum. However, momentum exchange is not continuous but is quantized in units of $2\pi/a$. The exchange process occurs every time there is an exchange of energy between a particle and the lattice. Then for optical transitions, $k_{f\perp} - k_{i\perp} = g$. The process is described in Figure 3.67c where the transition from an initial state $|i\rangle$ to a final state $|f\rangle$ is mediated by a wave vector $\mathbf{g}$. This process corresponds to a vertical transition in the *reduced Brillouin Scheme* where just one value of k is involved since the selection rule $\mathbf{g} = 2\pi/\mathbf{a}$, the Bloch wave function is the same for k and $k + g$ but does not affect E_f as seen in Figure 3.67c. For any excitation energy, just one transition can be observed. Varying the photon energy (synchrotron radiation), it is then possible to map the material band structure.

The transition probability from an initial state $|i\rangle$ to a final state $|f\rangle$ is described by the golden rule

$$w_{fi} = 2\pi/\hbar \; |\langle f| H_{int}|i\rangle| \; \delta(E_f - E_i - h\nu) \qquad 3.110$$

or using the wave vector k

$$w_{fi} = 2\pi/\hbar \langle f, k_f | H_{int}|i, k_i\rangle| \; \delta(E_f - E_i - h\nu) \qquad 3.111$$

Observe that the final state represents an excited level of the atom a. Following the three-step model, the photoelectron is emitted from the atom a into the material bulk. The final state $|f\rangle$ describes an excited state of the atom a immersed in a periodic potential (Bloch potential) if the material is crystalline. Then for this final level holds the selection rules on the momentum or, being $p = \hbar k$, on the wave vector: $k_f = k_i + g$.

If we assume that the M_{fi} matrix element is constant for a fixed excitation energy $h\nu$ as well as the density of the final states, from equation (3.111), it is possible to derive the photocurrent generated by the transition $|i\rangle \rightarrow |f\rangle$ as

$$N(E, h\nu) \propto |M_{fi}|^2 \sum_i \delta(E_f - E_i - h\nu) \, \delta(E - E_f + \phi)$$

$$N(E, h\nu) \propto |M_{fi}|^2 \, DOS(E_i) \, \delta(E - E_f + \phi) \qquad 3.112$$

The summation is extended to all the *one electron states* with energy E_i. The DOS(E_i) represents the density of occupied states at energy E_i meaning that the emitted photocurrent is proportional to the density of initial states DOS(E_i).

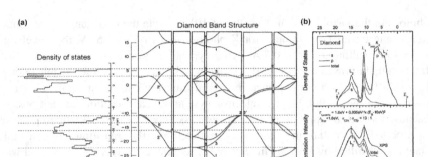

FIGURE 3.68   (a) Band structure and estimated density of states. (b) Calculated density of states of diamond (upper panel) and correspondence with the XPS valence band (lower panel) ([a] Reprinted with permission from [161]; [b] Reprinted with permission from [140]).

A qualitative correlation between valence band spectra and energy bands can be easily done considering that higher density of states for a defined energy corresponds to higher number of possible *k* states. This occurs when a E(k) band flattens. In Figure 3.68a, the band structure of diamond is correlated to the calculated density of states. Dashed lines are drawn in correspondence of flat regions of the band structure shown in Figure 3.68b.

3.10.4. VALENCE BAND ANALYSIS

As seen in the previous section, the valence band (VB) of a material reproduces the density of states of the material analyzed. If the band structure of that material is known, it can be interesting to fit the valence band to obtain at least qualitative information. In particular, the VB fit allows estimating the contribution of different orbitals to the DOS which can be useful to compare different materials. As an example, we can analyze the VB of graphite and diamond and retrieve the contribution of states deriving from *s* and *p* orbitals. This highlights the differences induced by the different hybridizations of the C-atoms resulting in the graphite and diamond crystalline systems. Figure 3.69 displays the VBs of diamond and graphite with its peak fitting.

The VB fit is computed adding a number of components in agreement with the *Band Structure* of crystalline graphite or crystalline diamond. After background subtraction using a Shirley curve, eight Gaussian components are added. The position of the components can be constrained in a small range around the correspondent theoretical position. After peak fitting, it is possible to measure the normalized spectral integral of the S and P parts as defined in ref. [112].

$$S \frac{\sum_{m=6}^{8} A_i}{\sum_{m=1}^{8} A_i} \qquad P = \frac{\sum_{m=1}^{3} A_i}{\sum_{m=1}^{8} A_i} \qquad\qquad 3.113$$

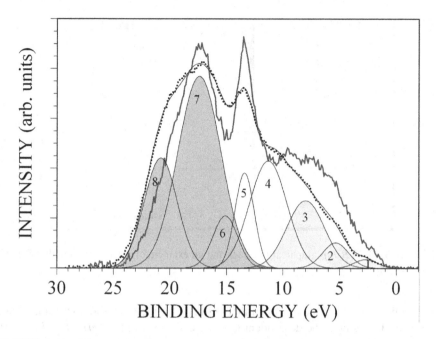

FIGURE 3.69 Diamond (dark solid line) and graphite (black dots) valence bands. The graphite valence band is fitted in agreement with calculated band structure.

The S and P parameters represent the set of colored components, respectively, at high and low BE shown in Figure 3.69. Despite the hybridization, orbitals still have memory of their original nature (unbonded atoms) if they were an s or a p orbital. Then the S parameter represents the contribution of the s-derived orbitals to the valence band while the P parameter mirrors that of the p-derived orbitals. As seen, in graphite and diamond, the S and P parameters does not interfere each other. It is possible to determine the excursion range of S and P from graphite, pure sp^2 system, to diamond, pure sp^3. On the base of these values, it is possible to estimate the abundance of sp^2 and sp^3 hybrids in mixed system as the amorphous carbons obtained by chemical or thermal processes or by plasma depositions.

3.10.5. ESTIMATION OF THE VALENCE BAND TOP AND FERMI LEVEL

The optical and electronic behavior of a material is governed, beside other, by the amplitude of the energy gap between valence and conduction bands (VB-CB gap). Photoelectron spectroscopies can only map states which are below the Fermi level. It is then not possible to measure the amplitude of the VB-CB gap. However, also the information relative the extension of the valence band toward the *Fermi Level* is important because it mirrors the conductive or insulating nature of the material. Because the DOS softly goes to zero when approaching the *Fermi Level* the correspondent trend of the VB smoothly goes to zero. This makes the identification of the VB edge, the *VB Top*, not immediate. *RxpsG* provides three different options to estimate the VB Top.

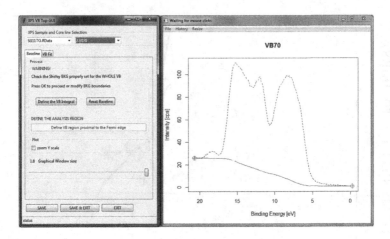

FIGURE 3.70   *VB Top* graphical interface and an example of valence band with a Shirley background.

The first method is the one commonly utilized for the estimation, which is based on linear fitting both the descendent tail of the VB toward the *Fermi Level* and the background level. The two fit will cross in a point which is considered as an estimate of the *VB Top*.

Figure 3.70 shows the *VB Top* GUI where, as usual, in the upper part the user can select the *XPS Sample* and the VB spectrum. To begin the analysis, the user is required to define the extension of the VB by adjusting the two red markers at the edges of the VB. Automatically the Shirley background is updated (Figure 3.70). The button *Reset Baseline* allows re-defining the VB edges. The user must be very careful to consider possible presence of spectral component at high BE (for example, in the present case, the contribution of oxygen 2s). Once the markers are in the correct position (as in the figure), the VB integral must be computed pressing the button *Define the VB Integral*.

This activates the evaluation of the VB integral and enables the definition of the VB region proximal to the *Fermi Edge*. Just clicking with the mouse, the analysis region is restricted to a region extending just few electron volts below the *Fermi Level* as illustrated in Figure 3.71. This increases the energy resolution and makes operations easier. Pressing the button *Define the VB Region proximal to the Fermi Edge*, the section relative to the VB fitting is activated.

The first method is the *Linear Fit*. The user must define two points on the descending tail of the VB and two additional points on the background. Pressing the button *Fit*, two linear fits through these points are computed as illustrated in Figure 3.72. The point where the two linear fit cross is assumed to be the *VB Top*. The exact value is obtained by pressing the button *Estimate the VB Top*. The other buttons are used to reset and restart the analysis.

The second option is the *Non-Linear Fit* of the VB. In this case, a series of Gaussian components is used to fit the VB. The Gaussian components are added just left clicking with the mouse on the desired position where a cross is visualized and pressing

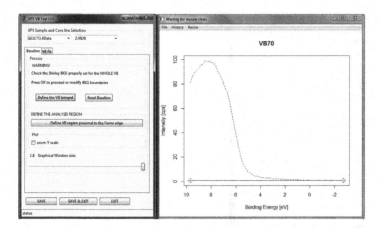

FIGURE 3.71   Definition of the VB region proximal to the Fermi Edge.

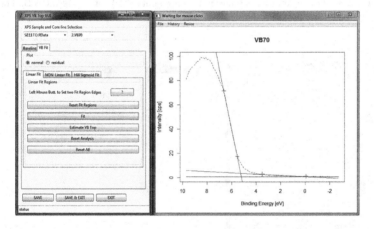

FIGURE 3.72   Estimation of the *VB Top* via linear fit. Observe the two couples of points indicated by the crosses and the relative linear fits. The common point of the two lines is assumed as an estimate of the *VB Top*.

the button *Add Fit Component* (Figure 3.73). It is not important that the number of components used neither their position nor FWHM after fitting. Important is that the best fit must perfectly describe the VB spectrum as illustrated in Figure 3.74.

The *Fit* button is used to obtain the best fit and the button *Estimate the VB Top* to compute the *VB Top* position as shown in Figure 3.74. In this option, the position of the *VB Top* is determined using a threshold defined from the VB integral computed at beginning of the procedure.

As in the previous case, the *Reset Analysis* and *Reset All* buttons are used to restart the analysis.

The last option is a VB fit using a Hill Sigmoid. In this case, the VB proximal to the Fermi Edge must be carefully selected to allow the sigmoid fitting.

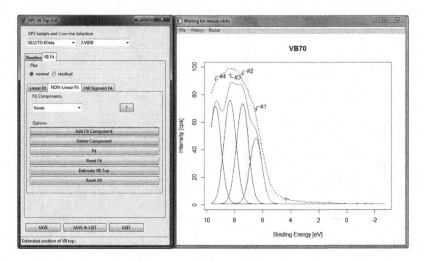

FIGURE 3.73   Definition of the Gaussian components used for fitting the VB. The cross defined with a left mouse click defines the position where to add an additional fit component.

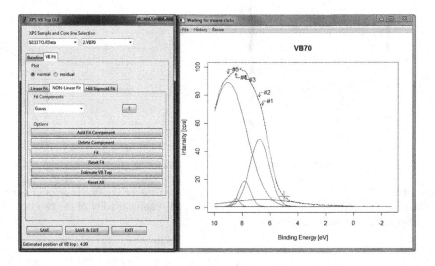

FIGURE 3.74   Result of the VB fit and definition of the *VB Top*.

Just clicking on the graph, the user must define the three points needed to add a Hill Sigmoid, the maximum and minimum values and that of the inflection point, which will be used to fit the VB (see figure 3.75). *Add Hill Sigmoid* and *Fit* buttons are used to add the curve and perform the best fit. The *VB Top* position is obtained by pressing the button *Estimate the VB Top*. The result is shown in Figure 3.76.

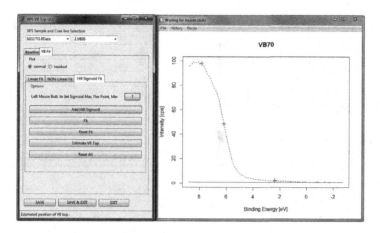

FIGURE 3.75 Definition of the correct VB region proximal to the Fermi Edge and of the three points needed for the Hill Sigmoid.

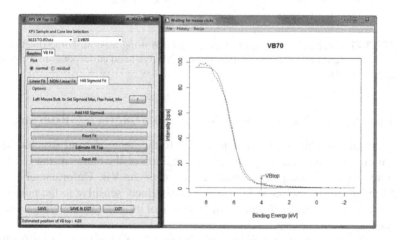

FIGURE 3.76 Hill Sigmoid VB fit and definition of the *VB Top*.

In addition to the possibility to estimate the *VB Top*, it is possible to estimate the position of the *Fermi Level* using the valence band of noble metals. Generally, silver or gold are used to calibrate the instrument by selecting the position of the *Fermi Level* as zero of the instrument binding energy scale. To compute the position of the Fermi Level, open the *Fermi Edge Estimation* option under the *Analysis* main menu and, similarly to what done in the VB Top procedure, define the step-like region of the VB by choosing the option *Define the VB Top Region* (see Figure 3.77). Once this VB portion is identified, select *Add Fermi Dirac Function* and then Fit the spectrum. The *Fermi Level position* is obtained by pressing the correspondent button.

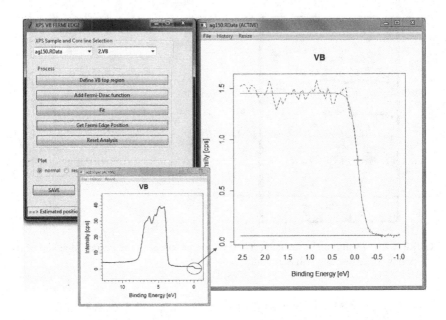

FIGURE 3.77   Step-like trend of the VB of Ag fitted using a Fermi-Dirac function. The position of the Fermi Edge appears at the bottom of the GUI and corresponds to the position of the cross overlapped to the VB fit.

3.10.6.   SECOND DERIVATIVE OF THE VALENCE BAND SPECTRA AND BAND STRUCTURE

The procedure to obtain information relative to the material band structure is the same utilized for the Auger spectrum. The position of the bands of the electronic structure is gained using the second derivative of the VB. Unfortunately, as for the Auger spectrum, the second derivative is extremely sensitive to the noise. This requires a careful spectral cleaning with rejection of the noise by applying a suitable filtering.

The user can increase step by step the power of the filter by saving the filtered spectrum using the smoothing test option and controlling the effect computing the second derivative of the VB.

Figure 3.78 illustrates the procedure. Both the *Smoothing* and *Differentiate* options are active. First, the VB is filtered and saved as a *Smoothing Test*. In Figure 3.78, this spectrum is indicated as ST.VB. Then the differentiation is applied to ST.VB. The negative of the second derivative is obtained by marking the *Negative* option, while *Amplify differentiated Data* is applied to amplify the oscillation of the second derivative otherwise hardly visible. In the example of Figure 3.78, the procedure is applied to the VB of HOPG.

As shown in [145], the oscillations of the negative of the second derivative of the VB correspond to the various components of the band structure of graphite. The result shown in Figure 3.78 is obtained applying an FIR filter 80 coefficients

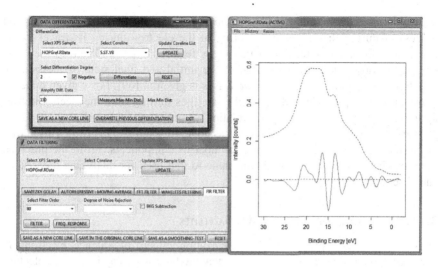

FIGURE 3.78 *Smoothing* option is applied to clean the spectrum (dashed line) while *Differentiate* is applied to compute the negative of the second derivative (solid line) shown on the right. The oscillations identify the components of the band structure.

and *Degree of Noise Rejection* = 18. The user can compare the results with those of ref. [145].

3.11. ENERGY LOSS FEATURES

In Section 3.7.9, we already presented the analysis of the core-lines including the fit of loss features.

The intrinsic part of the spectrum is commonly obtained by subtracting a background which represents the extrinsic contribution due to inelastic scattering processes of electrons deriving from the parent core-line. Besides asymmetric tails observed in conducting samples which account for a continuum of losses at low kinetic energies, additional source of losses are due to coupling of the photoelectron with the atom valence electrons or with electrons of the bulk atoms. In some cases, as it happens for transition metals, the loss features are part of the core-line spectrum and must be included in the peak fitting. Furthermore, loss features as *shake-up, shake-off* losses, carry chemical information because they appear only when specific bonds are formed [181], (see the case of CuO presented in Section 3.6.4 and more in general the case of transition metal oxides [118, 181–184] and rare earth [185–187]). In *shake-up, shake-off* processes, valence electrons are involved. In particular, as described in Section 3.6, the photoelectron interacts with the charge distribution of the atom releasing part of its energy to the valence electrons. The spectrum is then composed by a main peak and a number of additional lines, the satellites, representing the excited states. Commonly, loss features are referred to as *intrinsic* and *extrinsic* occurring, respectively, in the photoemission process or during the inelastic scattering of the electron during its travel to the material surface. The intensity of these loss features is proportional to the overlap of the wave functions associated to the initial and final state of the system. A detailed description of the *intrinsic* losses is given in [69]. For a given material in principle, it is possible to describe the satellites starting from the electronic configuration and, in the sudden approximation, trying to model the interaction of the photoelectron with the valence electrons as done, for example, in [188]. This allows fitting the satellites using the correct number of components. From the practical point of view, the fit procedure is the same to that used for analyzing a core-line and described in Section 3.7.

3.11.1. ANALYSIS OF THE SHAKE-UP SATELLITES

Shake-up satellites are produced by relaxation effects of valence electrons in response to the creation of a core-hole so that the final state may involve an electron in an excited bound state. These structures are observed in some alkali halides, small molecules, transition metals, and rare earth [184–186]. These structures were also observed in organic materials in a range extending till −20 eV below the core-line in particular in aromatic polymers. In this respect, polystyrene is a good example because possessing a pendant aromatic ring which is almost independent of the backbone thus assisting $\pi \rightarrow \pi^*$ transition. In polystyrene, the overall *shake-up* intensity is ~ 10% of that of the main peak. Figure 3.79a compares the C 1s of polypropylene (PP), polymethylmethacrylate (PMMA), polyethyleneterephthalate (PET), and polystyrene (PS).

As seen in Figure 3.79a, the loss components at high BE are observed only in polymers containing aromatic bonds. For this reason, the loss features of benzene are taken as a model. In gas phase, the benzene molecules display a well resolved loss structure with four principal components (see ref. [15]). Moving from low to high BE, the correspondent transitions are respectively the $2b_1 \rightarrow 3b_1$, the $1a_2 \rightarrow 2a_2$ which is the more intense, the $1b_1 \rightarrow 3b_1$, and the $1b_1 \rightarrow 4b_1$ shown in Figure 3.79d. Considering the chemical structure of PS with a pendant aromatic ring, the loss features are considered to be similar to those of benzene. Consequently the fit is made using four main components representing the same transitions. In Figure 3.79b, the loss structure of PS is fitted using a linear background and four Gaussian components. Similar fit is done on the C1s *shake-up* structures of PET [14]. Also PET contains an aromatic ring but differently form PS, the ring is bonded to two oxygen atoms. This induces small differences on the loss features mainly affecting the spectral intensity. Also in this case, following ref. [14], the fit can be performed using a linear background and four Gaussian components.

From the operative point of view, the linear background and the fitting components can be added using the *Analysis* option while *Fit Constraints* can be used to set the limits of the variability for the various fit parameters. As already observed, Section 3.7 describes these two options in detail providing all the needed information.

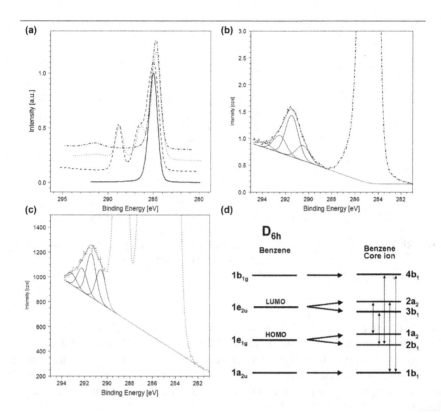

FIGURE 3.79 (a) Comparison of C1s of different polymers: PP (solid line), PMMA (dashed line), PET (dotted line), and PS (dash-dot line). Only PET and PS containing aromatic groups show *shake-up* features. (b) Fit of the PS *shake-up* spectrum using a linear background and Gaussian components. (c) Fit of the PET *shake-up* spectrum in the same conditions. (d) Electronic transitions of the benzene molecules in the gas phase.

3.11.2. PLASMON LOSSES AND STRUCTURAL INFORMATION

As seen in Section 3.65, photoelectrons generated somewhere in the bulk and moving toward the surface may excite collective oscillation of valence electrons, the *plasmons*, either in the bulk or at the solid surface. The plasmon loss features extend on the high BE side of the core-line, and are characterized by broad features composed by bulk and surface components. Owing to the collective nature of plasma oscillations, a complex dielectric function $\varepsilon(h\omega)$ is used to describe the loss spectra [189]. In this respect, the X-ray Photoelectron Spectra provide the electronic properties of the solid giving access to its dielectric function over a wide energy range. On the other side, the dielectric function is utilized to describe the allotropic forms of the material which can be compared with theoretical models or optical data [190]. This explains the sensitivity of the *plasmon* losses to the different material structures in crystalline as well as in amorphous forms [191]. A good example is the carbon materials which

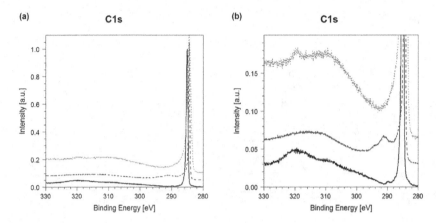

FIGURE 3.80   (a) C 1s of diamond (solid line), HOPG (dashed line), and amorphous graphite (dotted line). (b) The spectra are plotted on an expanded intensity scale to better see the different features of the *plasmon* losses.

can be produced in crystals as highly oriented crystalline graphite, and diamond or disordered forms as amorphous graphite or films possessing different abundances of sp^2, sp^3 hybrids depending on the synthesis conditions. C1s extended spectra acquired on diamond, highly oriented pyrolytic graphite, and amorphous graphite are shown in Figure 3.80a, b. Spectra presented on an expanded scale to better appreciate the loss features.

As it can be seen, different bulk structures produce different loss structures. In crystalline systems, the *plasmons* display a more structured spectrum while in amorphous systems they appear as a broad feature-less oscillation. Generally the description of the graphite *plasmon* losses assumes that excitation of π and σ electrons are independent processes. For graphite, this hypothesis is true if $\pi \rightarrow \pi^*$ and $\sigma \rightarrow \sigma^*$ excitations occur in the plain. In a general view, however, also the out-of-plain interband transitions $\sigma \rightarrow \pi^*$ and $\pi \rightarrow \sigma^*$ must be considered. The graphite loss spectrum is formed by two main structures. Near the zero-loss C 1s peak, at ~ 6 eV falls the $\pi \rightarrow \pi^*$ loss, a pure transition involving only π electrons. At loss energies higher than 10 eV, a more broad structure derives from all π and σ valence electrons. Following the interpretation given in [192], going from lower to higher energy loss values, this broad structure may be explained by the sequence of transitions $\sigma \rightarrow \sigma^*$ at ~12.5 eV involving σ electrons from deriving from p-like orbitals [192], at ~15 eV is found the $\sigma - \pi$ transition while the component at ~20 eV is ascribed to the out of plane *plasmon* [193]. The $\sigma + \pi$ transition describes the contribution at ~26 eV and is the more intense since all the valence electrons are involved in this *plasmon* excitation. At higher loss energy are involved $\sigma \rightarrow \sigma^*$ transitions deriving from deep s-like orbitals of the valence band [192]. This description was used to fit the *plasmon* loss spectrum as reported in Figure 3.81a.

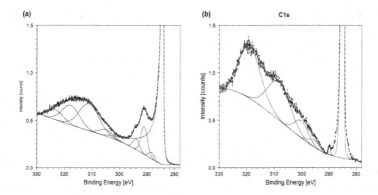

FIGURE 3.81   (a) Fit of the C1s and of the *plasmon* losses of HOPG. (b) Fit of the C1s and *plasmon* losses of diamond.

In the case of diamond, the first loss component is found at ~ 5 eV. Remembering that the energy gap of diamond is 5.45 eV, there should not be any loss component below this energy. Losses found at lower energy likely derive from possible π states deriving from surface reconstruction. In addition to defects and presence of graphitic component on the diamond surface, absorption spectra [194] can explain the σ → σ* also in the case of diamond explaining the presence of the components at ~11 eV [195]. Finally, at ~ 24 eV and 34 eV fall two main components dominating the loss spectrum. They are due to σ valence electrons and are commonly described as bulk and surface *plasmon* respectively. The correspondent fit of the diamond energy loss spectrum is shown in Figure 3.81b.

In general, the analysis, the number of fit components, and the selection of their positions are made referring to appropriate works in literature providing the correct interpretation of the energy losses. In these specific cases, the HOPG spectrum was fitted using a 4.P Tougaard baseline, the zero-loss peak was described with a Doniach-Sunjic function, while Gaussian components were used for the loss structures. Concerning diamond, similar to HOPG background subtraction should be made using the Tougaard baseline. However, due to the limited energy range, this baseline crosses the spectrum at ~290 eV in the region between the zero-loss peak and the main *plasmon*. For this reason, a spline baseline was used to do the background subtraction. The zero-loss C1s peak was fitted using an asymmetric Gauss-Voigt function while Gaussian components were used for the *plasmon* losses.

REFERENCES

[1]   W.D. Kaplan, D. Chatain, P. Wynblatt, C. Carter, A review of wetting versus adsorption, complexions, and related phenomena: The Rosetta stone of wetting, J. Mater. Sci., 48 (2013) 5681–5717.

[2]   M. Callewaert, J.-F. Gohy, C.C. Dupont-Gillain, L. Boulange-Petermann, P.G. Roux-het, Surface morphology and wetting properties of surfaces coated with an amphiphilic diblock copolymer, Surf. Sci., 575 (2005) 125–135.

[3]   K.L. Mittal, Contact Angle, Wettability and Adhesion, CRC Press, Boca Raton, FL, 2018.

[4]   G.A. Somorjai, Y. Li, Impact of surface chemistry, Proc. Natl. Acad. Sci., 108 (2011) 917–924.

[5]   D. Landolt, Corrosion and Surface Chemistry of Metals, CRC Press, Boca Raton, FL, 2007.

[6]   E. McCafferty, Surface Chemistry of Aqueous Corrosion Processes, SpringerBriefs in Materials, Springer, Cham; Heidelberg; New York; Dordrecht; London, 2015.

[7]   G.A. Somorjai, Y. Li, Introduction to Surface Chemistry and Catalysis, Wiley, Hoboken, NJ, 2010.

[8]   K.W. Kolasinski, Surface Science Foundations of Catalysis and Nanoscience, Wiley, Chichester, 2008.

[9]   S. Kamble, S. Agrawal, S. Cherumukkil, V. Sharma, R.V. Jasra, P. Munshi, Revisiting zeta potential, the key feature of interfacial phenomena, with applications and recent advancements, ChemistrySelect, 7 (2022) e202103084.

[10]   Z. Kolska, Z. Makajova, K. Kolarova, N. Kasalkova-Slepickova, S. Trostova, A. Reznickova, J. Siegel, V. Svorcik, Electrokinetic Potential and Other Surface Properties of Polymer Foils and Their Modifications, in: Polymer Science, F. Yilmaz Ed., IntechOpen, https://www.intechopen.com/predownload/42101, 2013.

[11]   J. Rech, H. Hamdi, S. Valette, Chapter 3 – Workpiece Surface Integrity, in: Machining: Fundamentals and Recent Advances, J.P. Davim Ed., Springer-Verlag, London, 2008.

[12]   J.F. Watts, J. Wolstenholme, An Introduction to Surface Analysis by XPS and AES, Wiley, Chichester, 2003.

[13]   P. Van der Heide, X-ray Photoelectron Spectroscopy—An Introduction to Principles and Practices, John Wiley & Sons Inc., Hoboken, NJ, 2012.

[14]   G. Beamson, D. Briggs, High Resolution XPS of Organic Polymers: The Scienta ESCA300 Database, Surface Spectra, Chichester, 1992.

[15]   D. Briggs, Surface Analysis of Polymers by XPS and Static SIMS, D.R. Clarke, S. Suresh, I.M. Ward Eds, Cambridge University Press, Cambridge, 1998.

[16]   R.T. Haasch, T.-Y. Lee, D. Gall, J.E. Greene, I. Petrov, Epitaxial ScN(001) grown and analyzed in situ by XPS and UPS I analysis of As-deposited layers, Surf. Sci. Spectra, 7 (2000) 169.

[17]   R.T. Haasch, T.-Y. Lee, D. Gall, J.E. Greene, I. Petrov, Epitaxial TiN(001) grown and analyzed in situ by XPS and UPS I analysis of As-deposited layers, Surf. Sci. Spectra, 7 (2000) 193.

[18]   R.T. Haasch, T.-Y. Lee, D. Gall, J.E. Greene, I. Petrov, Epitaxial VN(001) grown and analyzed in situ by XPS and UPS I analysis of As-deposited layers, Surf. Sci. Spectra, 7 (2000) 221.

[19]   R.T. Haasch, T.-Y. Lee, D. Gall, J.E. Greene, I. Petrov, Epitaxial CrN(001) grown and analyzed in situ by XPS and UPS I analysis of As-deposited layers, Surf. Sci. Spectra, 7 (2000) 250.

[20]   S. Hufner, Unfilled InnerShells: Transition Metals and Compounds, in: Photoemission in Solids, II edition by L. Ley, M. Cardona Eds., Springer, Berlin; Heidelberg; New York, 1979: pp. 173–216.

[21]   S. Hufner, Photoelectron Spectroscopy, 3rd Edition, Springer, Berlin, 2003.

[22]   L. Kover, Chemical Effects in XPS, in: Surface Analysis by Auger and Photoelectron Spectroscopy, D. Briggs, J.T. Grant Eds., IMP Publications, Chichester, 2003: pp. 421–464.

[23] W.J. Mortier, Electronegativity Equalization and Its Applications, in: Electronegativity. Structure and Bonding, K.D. Sen, K.D. Jorgensen Eds., Springer, Berlin; Heidelberg, 1987.

[24] A.V. Naumkin, A. Kraut-Vass, S.W. Gaarenstroom, C.J. Powell, NIST X-ray Photoelectron Spectroscopy Database—NIST Standard Reference Database 20, Version 41, https://srdata.nist.gov/xps/Default.aspx, 2012.

[25] J.F. Moulder, W.F. Stickle, P.E. Sobol, K.D. Bomben, Handbook of X-ray Photoelectron Spectroscopy: A Reference Book of Standard Spectra for Identification and Interpretation of XPS Data, Physical Electronics, Eden Prairie, MN, 1995.

[26] R. Bonoit, LaSurface.com, http://www.lasurface.com/database/elementxps.php, n.d.

[27] N. Ikeo, Y. Iijima, N. Niimura, M. Sigematsu, T. Tazawa, S. Matsumoto, K. Kojima, Y. Nagasawa, Handbook of X-ray Photoelectron Spectroscopy, JEOL, Akishima, 1991.

[28] T. Grant, Databases, in: Surface Analysis by Auger and Photoelectron Spectroscopy, D. Briggs, J.T. Grant Eds., IMP Publications, Chichester, 2003: pp. 869–873.

[29] Y.B. Band, Y. Avishai, 4–Spin, in: Quantum Mechanics with Applications to Nanotechnology and Information Science, Academic Press-Elsevier, Oxford, 2013: pp. 159–192.

[30] M.P. Seah, W.A. Dench, Quantitative electron spectroscopy of surfaces: A standard data base for electron inelastic mean free paths in solids, Surf. Interface Anal., 1 (1979) 2–11.

[31] C.J. Powell, H. Shinotsuka, S. Tanuma, D.R. Penn, Calculations of electron inelastic mean free paths XII Data for 42 inorganic compounds over the 50 eV to 200 keV range with the full Penn algorithm, Surf. Interface Anal., 51 (2019) 427–457.

[32] C.J. Powel, Practical guide for inelastic mean free paths, effective attenuation lengths, mean escape depths, and information depths in x-ray photoelectron spectroscopy, J. Vac. Sci. Technol. A, 38 (2020) 023209.

[33] M.P. Seah, An accurate and simple universal curve for the energy-dependent electron inelastic mean free path, Surf. Interface Anal., 44 (2012) 497–503.

[34] A. Jablonski, C.J. Powel, Effective attenuation lengths for different quantitative applications of X-ray photoelectron spectroscopy, J. Phys. Chem. Ref. Data, 49 (2020) 033102.

[35] S. Tanuma, C.J. Powel, D.R. Penn, Calculations of electron inelastic mean free paths V Data for 14 organic compounds over the 50–2000 eV range, Surf. Interface Anal., 21 (1994) 165–176.

[36] S. Tanuma, C.J. Powel, D.R. Penn, Calculations of electron inelastic mean free paths II Data for 27 elements over the 50–2000 eV range, Surf. Interface Anal., 17 (1991) 911–926.

[37] P.J. Cumpson, M.P. Seah, Elastic scattering corrections in AES II Estimating attenuation lengths and conditions required for their valid use in overlayer/substrate experiments, Surf. Interface Anal., 25 (1997) 430–446.

[38] C.J. Powel, A. Jablonski, NIST Electron Inelastic-Mean-Free-Path Database: Version 12, https://www.nist.gov/srd/nist-standard-reference-database-71, 2010.

[39] S. Hofmann, Chapter 4 – Quantitative Analysis (Data Evaluation), in: Auger- and X-Ray Photoelectron Spectroscopy in Materials Science: A User-Oriented Guide, G. Ertl, H. Luth, D.L. Mills, Series Eds, Springer-Verlag, Berlin; Heidelberg, 2013.

[40] W.S.M. Werner, S. Tanuma, Surface Sensitivity, in: Surface Analysis by Auger and Photoelectron Spectroscopy, IMP Publications, Chichester, 2003: pp. 235–294.

[41] M.P. Seah, Quantification of AES and XPS, in: Practical Surface Analysis Vol. 1 Auger and X-ray Photoelectron Spectroscopy, D. Briggs, M.P. Seah Eds., Johns Wiley & Sons, Chichester, 1990: pp. 206–251.

[42] C.S. Fadley, X-ray photoelectron spectroscopy: Progress and perspectives, J. Electron Spectrosc. Relat. Phenom., 178–179 (2010) 2–32.

[43] R.F. Reilman, A. Msezane, S.T. Manson, Relative intensities in photoelectron spectroscopy of atoms and molecules, J. Electron Spectrosc. Relat. Phenom., 8 (1976) 389–394.

[44] J.H. Scofield, Hartree-Slater subshell photoionization cross-sections at 1254 and 1487 eV, J. Electron Spectrosc. Relat. Phenom., 8 (1976) 129–137.

[45] C.D. Wagner, L.E. Davis, M.V. Zeller, J.A. Taylor, R.H. Raymond, L.H. Gale, Empirical atomic sensitivity factors for quantitative analysis by electron spectroscopy for chemical analysis, Surf. Interface Anal., 3 (1981) 211–225.

[46] C.R. Brundle, B.V. Crist, X-ray photoelectron spectroscopy: A perspective on quantitation accuracy for composition analysis of homogeneous materials, J. Vac. Sci. Technol. A, 38 (2020) 041001.

[47] A. Shirley, High-resolution X-ray photoemission spectrum of the valence bands of gold, Phys. Rev. B, 5 (1972) 4709.

[48] J. Vegh, The Shirley-equivalent electron inelastic scattering cross-section function, Surf. Sci., 563 (2004) 183–190.

[49] D. Briggs, M.P. Seah, Practical Surface Analysis by Auger and Photoelectron Spectroscopies, John Wiley & Sons, Chichester, 1990.

[50] S. Tougaard, C. Jansson, Comparison of validity and consistency of methods for quantitative XPS peak analysis, Surf. Interface Anal., 20 (1993) 1013–1046.

[51] S. Tougaard, Universality classes of inelastic electron Scattering cross-sections, Surf. Interface Anal., 25 (1997) 137–154.

[52] S. Tougaard, Surface nanostructure determination by X-ray photoemission spectroscopy peak shape analysis, J. Vac. Sci. Technol. A, 14 (1996) 1415.

[53] S. Tougaard, Quantitative x-ray photoelectron spectroscopy: Simple algorithm to determine the amount of atoms in the outermost few nanometers, J. Vac. Sci. Technol. A, 21 (2003) 1081.

[54] S. Hajati, S. Coult, C. Bloomfield, S. Tougaard, XPS imaging of depth profiles and amount of substance based on Tougaard's algorithm, Surf. Sci., 600 (2006) 3015–3021.

[55] C.J. Powel, J.M. Conny, Evaluation of uncertainties in X-ray photoelectron spectroscopy intensities associated with different methods and procedures for background subtraction II Spectra for unmonochromated Al and Mg X-rays, Surf. Interface Anal., 41 (2009) 804–813.

[56] G.H. Major, N. Fairley, P.M.A. Sherwood, M.R. Linford, J. Terry, V. Fernandez, K. Artyushkova, Practical guide for curve fitting in X-ray photoelectron spectroscopy, J. Vac. Sci. Technol. A, 38 (2020) 061302.

[57] S. Hufner, G.K. Wertheim, Core-line asymmetries in the X-ray-photoemission spectra of metals, Phys. Rev. B, 11 (1975) 678.

[58] S. Hufner, G.K. Wertheim, Many-body line shape in X-ray photoemission from metals, Phys. Rev. Lett., 35 (1975) 53.

[59] P.H. Citrin, P. Eisenberger, D.R. Hamann, Phonon broadening of X-ray photoemission linewidths, Phys. Rev. Lett., 33 (1974) 965.

[60] P.M.A. Sherwood, Data Analysis in XPS and AES, in: Practical Surface Analysis, 2nd Edition, Wiley, Chichester, 1990: pp. 555–586.

[61] N. Fairley, XPS Lineshapes and Curve Fitting, in: Surface Analysis by Auger and Photo-Electron Spectroscopy, D. Briggs, J.T. Grant Eds., IMP Publications, Chichester, 2003: pp. 412–464.

[62] S. Doniach, M. Sunjic, Many-electron singularity in X-ray photoemission and X-ray line spectra from metals, J. Phys. C, 3 (1970) 285.

[63] G.K. Wertheim, P.H. Citrin, Fermi Surface Excitations in X-Ray Photoemission Line Shapes from Metals, in: Photoemission in Solids, M. Cardona, L. Ley Eds., Springer, Berlin; Heidelberg, 1978: pp. 197–236.

[64] K. Siegbahn, Electron spectroscopy—an outlook, J. Electron Spectrosc. Relat. Phenom., 5 (1974) 3–97.

[65] G.K. Wertheim, S. Hufner, H.J. Guggenheim, Systematics of core-electron exchange Splitting in 3d-group transition-metal compounds, Phys. Rev. B, 7 (1973) 556.

[66] S. Hufner, G.K. Wertheim, Core-electron splittings and hyperfine fields in transition-metal compounds, Phys. Rev. B, 7 (1973) 2333.

[67] R.L. Cohen, G.K. Wertheim, A. Rosencwaig, H.J. Guggenheim, Multiplet splitting of the 4s and 5s electrons of the rare earths, Phys. Rev. B, 5 (1972) 1037.

[68] F.R. McFeely, S.P. Kowalczyk, L. Ley, D.A. Shirley, Multiplet splittings of the 4s and 5s core levels in the rare earth metals, Phys. Lett. A, 49 (1974) 301.

[69] S. Hufner, Photoelectron Spectroscopy: Principles and Applications, 3rd Edition, Springer, Berlin, 2003.

[70] P.H. Citrin, T.D. Thomas, X-ray photoelectron spectroscopy of alkali halides, J. Chem. Phys., 57 (1972) 4446.

[71] J.H. Van Vleck, The Dirac vector model in complex spectra, Phys. Rev., 45 (1934) 405.

[72] P. Van der Heide, Multiplet splitting patterns exhibited by the first row transition metal oxides in X-ray photoelectron spectroscopy, J. Electron Spectrosc. Relat. Phenom., 164 (2008) 8.

[73] C.S. Fadley, Electron Spectroscopy: Theory, Techniques and Applications, C.R. Brundle, A.D. Baker Eds., Academic Press, London, New York, San Francisco, CA, 1978.

[74] A. Kotani, Y. Toyozawa, Photoelectron spectra of core electrons in metals with an incomplete shell, J. Phys. Soc. Jpn., 37 (1974) 912–919.

[75] M. Kurth, P.C.J. Graat, E.J. Mittemeijer, Determination of the intrinsic bulk and surface plasmon intensity of XPS spectra of magnesium, Appl. Surf. Sci., 220 (2003) 60–68.

[76] K. Levenberg, A method for the solution of certain non-linear problems in least squares, Q. Appl. Math., 2 (1944) 164–168.

[77] D. Marquardt, An algorithm for least-squares estimation of nonlinear parameters, SIAM J. Appl. Math., 11 (1963) 431–441.

[78] D. Bindel, D. Bindel, Spring 2016 Numerical Analysis (CS 4220) in Numerical Analysis: Linear and Nonlinear Problems (CS 4220/5223/MATH 4260), 2016.

[79] A. Croeze, L. Pittman, W. Reynolds, Solving the Non-Linear Least-Squares Problem with Gauss-Newton and Levenberg-Marquardt Methods, https://www.math.lsu.edu/system/files/MunozGroup1 - Paper.pdf, 2022.

[80] H.P. Gavin, The Levenberg-Marquardt Algorithm for Nonlinear Least Squares Curve-Fitting Problems, https://people.duke.edu/~hpgavin/ExperimentalSystems/lm.pdf, 2020.

[81] The PORT3 Library, http://www.netlib.org/port/, 2022.

[82] P.A. Fox, A.D. Hall, N.L. Schryer, The PORT Mathematical Subroutine Library, ACM Trans. Math. Softw., 4 (1978), 104–126.

[83] D.M. Gay, Computing Science Technical Report No. 153-Usage Summary for Selected Optimization Routines, AT&T Bell Laboratories, Murray Hill, NJ. https://math.mcmaster.ca/~bolker/misc/port.pdf, 1990.

[84] J.E. Dennis, D.M. Gay, R.E. Walsh, An Adaptive Nonlinear Least-Square Algorithm, ACM Trans. Math. Softw., 7(3) (1981), 348–368.

[85] J. Nelder, R. Mead, A simplex method for function minimization, Comput. J., 7 (1965) 308–313.

[86]   M.J.D. Powell, On search directions for minimization algorithms, Math. Program., 4 (1973) 193–201.

[87]   M.R. Hestenes, E. Stiefel, Methods of conjugate gradients for solving linear systems, J. Res. Natl. Bur. Stand., 49 (1952) 23.

[88]   C. Broyden, A class of methods for solving nonlinear simultaneous equations, Math. Comp., 19 (1965) 577–593.

[89]   C. Broyden, The convergence of a class of double rank minimization algorithms, IMA J. Appl. Math., 6 part I and II (1970) 76–90 and 222–231.

[90]   C.J.P. Belisle, Convergence theorems for a class of simulated annealing algorithms, J. Appl. Probab., 29 (1992) 885–895.

[91]   W.L. Price, A controlled random search procedure for global optimization, Comput. J., 20 (1977) 367–370.

[92]   W.L. Price, Global optimization by controlled random search, J. Optim. Theory Appl., 40 (1983) 333–348.

[93]   R.W. Schafer, What is a Savitzky-Golay filter? IEEE Signal Process. Mag., 28 (2011) 111–117.

[94]   R.W. Shafer, On the frequency-domain properties of the Sawitzky-Golay filters, Proceedings of the "Digital Signal Processing Workshop and IEEE Signal Processing Education Meeting (DSP/SPE)", Sedona, AZ, 4–7 January 2011: pp. 54–59.

[95]   M. Schmid, D. Rath, U. Diebold, Why and how Savitzky-Golay filters should be replaced, ACS Meas. Sci. Au, 2 (2022) 185–196.

[96]   G. Box, G.M. Jenkins, Time Series Analysis: Forecasting and Control, 3rd Edition, Prentice-Hall, Englewood Cliffs, NJ, 1994.

[97]   R. Shumway, D. Stoffer, Time Series Analysis and Its Applications: With R Examples, 3rd Edition, Springer, New York, 2010.

[98]   J.W. Cooley, J.W. Tukey, An algorithm for the machine calculation of complex Fourier series, Math. Comp., 19 (1965) 297–301.

[99]   G.P. Nason, Wavelet Methods in Statistics with R, 1st Edition, Springer Science, New York, 2008.

[100] C.S. Burrus, R.A. Gopinath, H. Guo, Introduction to Wavelets and Wavelet Transforms: A Primer, Prentice Hall, Upper Saddle River, NJ, 1998.

[101] A. Abbate, C.M. DeCusatis, P.K. Das, Wavelets and Subbands—Fundamentals and Applications, Birkheuser, Boston, MA, 2002.

[102] R. Polikar, The Wavelet Tutorial, https://cseweb.ucsd.edu/~baden/Doc/wavelets/polikar_wavelets.pdf, 2004.

[103] A. Proctor, P.M.A. Sherwood, Data analysis techniques in X-ray photoelectron spectroscopy, Anal. Chem., 54 (1982) 13–19.

[104] F. Holler, D.H. Burns, J.B. Callis, Direct use of second derivatives in curve-fitting procedures, Appl. Spectrosc., 43 (1989) 877.

[105] A.E. Pavlath, M.M. Millard, Analysis of X-ray photoelectron spectra through their even derivatives, Appl. Spectrosc., 33 (1979) 502.

[106] J.M. Conny, C.J. Powell, L.A. Currie, Standard test data for estimating peak parameter errors in X-ray photoelectron spectroscopy I peak binding energies, Surf. Interface Anal., 26 (1998) 939–956.

[107] L.S. Daley, M.M. Thompson, W.M. Proebsting, J. Postman, B. Jeong, Use of fourth-derivative visible spectroscopy of leaf lamina in plant germplasm characterization, Spectroscopy, 1 (1986) 28.

[108] J.S. Brown, S. Schoch, Spectral analysis of chlorophyll-protein complexes from higher plant chloroplasts, Biochim. Biophys. Acta, 636 (1981) 201–209.

[109] S. Tougaard, Energy loss in XPS: Fundamental processes and applications for quanti-fication, non-destructive depth profiling and 3D imaging, J. Electron Spectrosc. Relat. Phenom., 178–179 (2010) 128–153.

[110] F. Yubero, S. Tougaard, Quantification of plasmon excitations in core-level photoemis-sion, Phys. Rev. B, 71 (2005) 045414.

[111] M. Kurth, P.C.J. Graat, Quantitative analysis of the plasmon loss intensities in X-ray photoelectron spectra of magnesium, Surf. Interface Anal., 34 (2002) 220–224.

[112] G. Speranza, N. Laidani, Measurement of the relative abundance of sp2 and sp3 hybri-dised atoms in carbon based materials by XPS: A critical approach Part I, Diam. Relat. Mater., 13 (2004) 445–450.

[113] N. Moslemzadeh, G. Beamson, P. Tsakiropoulos, J.F. Watts, S.R. Haines, P. Weight-man, The 1s XPS spectra of the 3d transition metals from scandium to cobalt, J. Elec-tron Spectrosc. Relat. Phenom., 152 (2006) 129–133.

[114] A.P. Grosvenor, B.A. Kobe, M.C. Biesinger, N.S. McIntyre, Investigation of multiplet splitting of Fe 2p XPS spectra and bonding in iron compounds, Surf. Interface Anal., 36 (2004) 1564–1574.

[115] R. Li, C. Sun, J. Liu, Q. Zhen, Sulfur-doped CoFe2 O4 nanopowders for enhanced visible-light photocatalytic activity and magnetic properties, RSC Adv., 7 (2017) 50546.

[116] A.P. Grosvenor, M.C. Biesinger, R.S.C. Smart, N.S. McIntyre, New interpretations of XPS spectra of nickel metal and oxides, Surf. Sci., 600 (2006) 1771–1779.

[117] M.C. Biesinger, Advanced analysis of copper X-ray photoelectron spectra, Surf. Inter-face Anal., 49 (2017) 1325–1334.

[118] M.C. Biesinger, B.P. Payne, L.W.M. Lau, A. Gerson, R.S.C. Smart, X-ray photoelectron spectroscopic chemical state quantification of mixed nickel metal, oxide and hydroxide systems, Surf. Interface Anal., 41 (2009) 324–332.

[119] L. Meitner, Uber die Entstehung der β-Strahl-Spektren radioaktiver Substanzen, Z. Physik, 9 (1922) 131.

[120] P. Auger, Sur l'feffet photoelectrique compose, J. Phys. Radium, 6 (1925) 205.

[121] J.J. Lander, Auger peaks in the energy spectra of secondary electrons from various, Phys. Rev., 91 (1953) 1382.

[122] J.W. Smith, R.J. Saykally, Soft X-ray absorption spectroscopy of liquids and solutions, Chem. Rev., 117 (2017) 13909–13934.

[123] J. Wolstenholme, Auger Electron Spectroscopy: Practical Application to Materials Analysis and Characterization of Surfaces, Interfaces, and Thin Films, Momentum Press, New York, 2015.

[124] D. Briggs, J.T. Grant, Surface Analysis by Auger and X-ray Photoelectron Spectrosco-pies, 1st Edition, IM Publications and Surface Spectra, Trowbridge, 2003.

[125] Chemistry, Periodic Trend in Difference of Energy between the S and P Orbitals, https://chemistry.stackexchange.com/questions/16522/periodic-trend-in-difference-of-energy-between-the-s-and-p-orbitals, 2014.

[126] Eurofins EAG-Laboratories, Auger Electron Energies, https://www.eag.com/resources/tutorials/auger-tutorial-theory/, 2015.

[127] J.E. Houston, Local Electronic Structure Information in Auger Electron Spectroscopy: Solid Surfaces, in: Treatise on Materials Science and Technology—Auger Electron Spectroscopy, C.L. Briant, R.P. Messmer Eds., Academic Press, San Diego, CA, 1988: pp. 65–110.

[128] G.E. McGuire, Auger Electron Spectroscopy Reference Manual—A Book of Standard Spectra for Identification and Interpretation of Auger Electron Spectroscopy Data, Springer, New York, 1979.

[129] K.D. Childs, Handbook of Auger Electron Spectroscopy—A Reference Book of Standard Data for Identification and Interpretation of Auger Electron Spectroscopy Data, Physical Electronics, Eden Prairie, MN, 1995.

[130] C.D. Wagner, Electron Spectroscopy, Proceedings of International Conference—Asilomar, Pacific Grove, CA, 7–10 September, 1971, D.A. Shirley Ed., Amsterdam, 1972: p. 861.

[131] C.D. Wagner, Auger lines in X-ray photoelectron spectrometry, Anal. Chem., 44 (1972) 967–973.

[132] C.D. Wagner, Chemical shifts of Auger lines, and the Auger parameter, Faraday discuss. Chem. Soc., 60 (1975) 291–300.

[133] G. Moretti, Auger parameter and Wagner plot in the characterization of chemical states by X-ray photoelectron spectroscopy: A review, J. Electron Spectrosc. Relat. Phenom., 95 (1998) 95–114.

[134] M. Satta, G. Moretti, Auger parameters and Wagner plots, J. Electron Spectrosc. Relat. Phenom., 178–179 (2010) 123–127.

[135] Y. Furukawa, Y. Nagatsuka, Y. Nagasawa, S. Fukushima, M. Yoshitake, A. Tanaka, Practical methods for detecting peaks in Auger electron spectroscopy and X-ray photoelectron spectroscopy, J. Surf. Anal., 14 (2008) 225–252.

[136] J.C. Lascovich, I.A. Santoni, Study of the occupied electronic density of states of carbon samples by using second derivative carbon WV Auger spectra, Appl. Surf. Sci., 106 (1996) 245–253.

[137] J.P. Coad, J.C. Riviere, The LMM Auger spectra of some transition metals of the first series, Z. Physik, 244 (1971) 19–30.

[138] P.G. Lurie, J.M. Wilson, The diamond surface: II Secondary electron emission, Surf. Sci., 65 (1977) 476–498.

[139] A. Bianconi, S.B.M. Hagstrom, R.Z. Bachrach, Photoemission studies of graphite high-energy conduction-band and valence-band states using soft-X-ray synchrotron radiation excitation, Phys. Rev. B, 16 (1977) 5534.

[140] J. Schafer, J. Ristein, R. Graupner, L. Ley, U. Stephan, T. Frauenheim, V.S. Veersamy, G.A.J. Amaratunga, M. Weiler, H. Ehrhardt, Photoemission study of amorphous carbon modifications and comparison with calculated densities of states, Phys. Rev. B, 53 (1996) 7762.

[141] M. Cini, Theory of Auger XVV spectra of solids: Many body effects in incomplete filled bands, Surf. Sci., 87 (1979) 483–500.

[142] M.A. Smith, L.L. Levenson, Valence-band information from the Auger K VV spectrum of graphite, Phys. Rev. B, 16 (1977) 2973.

[143] W.M. Mularie, W.T. Peria, Deconvolution techniques in Auger spectroscopy, Surf. Sci., 26 (1971) 125–141.

[144] J.E. Houston, J.W. Rogers, R.R. Rye, F.L. Huston, D.D. Ramaker, Relationship between the Auger line shape and the electronic properties of graphite, Phys. Rev. B, 34 (1986) 1215.

[145] L. Calliari, G. Speranza, J.C. Lascovich, I.A. Santoni, The graphite core–valence–valence Auger spectrum, Surf. Sci., 501 (2002) 253–260.

[146] E. Perfetto, M. Cini, S. Ugenti, P. Castrucci, M. Scarselli, M. De Crescenzi, F. Rosei, M.A. El Khakani, Electronic correlations in graphite and carbon nanotubes from Auger spectroscopy, Phys. Rev. B, 76 (2007) 233408.

[147] W.A. Coghlan, R.E. Clausing, A description of a catalog of calculated Auger transition for the elements, Surf. Sci., 33 (1972) 411–413.

[148] J.A.D. Matthew, A modified Z/(Z+1) approximation for free-atom Auger spectra, J. Phys. B, 10 (1977) 783.

[149] D. Chattarji, The Theory of Auger Transitions, 1st Edition, Academic Press, London; New York; San Francisco, CA, 1976.

[150] L.E. Davis, Handbook of Auger Electron Spectroscopy: A Reference Book of Standard Data for Identification and Interpretation of Auger Electron Spectroscopy Data, 3rd Edition, Physical Electronics, Eden Prairie, MN, 1996.

[151] A.A. Galuska, H.N. Madden, R.E. Allred, Electron spectroscopy of graphite, graphite oxide and amorphous carbon, Appl. Surf. Sci., 32 (1988) 253–272.

[152] J.C. Lascovich, R. Giorgi, S. Scaglione, Evaluation of the sp2/sp3 ratio in amorphous carbon structure by XPS and XAES, Appl. Surf. Sci., 47 (1991) 17–21.

[153] P.H. Van Cittert, Zum Einfluss der Spaltbreite auf die Intensitatswerteilung in Spektrallinien II, Z. Physik, 69 (1931) 298.

[154] C. Xu, I. Aissaoui, S. Jacquey, Algebraic analysis of the Van Cittert iterative method of deconvolution with a general relaxation factor, J. Opt. Soc. Am. A, 11 (1994) 2804.

[155] P. Bandzuch, M. Morhac, T. KriStiak, Study of the Van Cittert and Gold iterative methods of deconvolution and their application in the deconvolution of experimental spectra of positron annihilation, Nucl. Instrum. Methods Phys. Res. A, 384 (1997) 506–515.

[156] C.S. Fadley, X-ray photoelectron spectroscopy: From origins to future directions, Nucl. Instrum. Methods Phys. Res. A, 601 (2009) 8–31.

[157] L. Chen, R. Batra, R. Ranganathan, G. Sotzing, Y. Cao, R. Ramprasad, Electronic structure of polymer dielectrics: The role of chemical and morphological complexity, Chem. Mater., 30 (2018) 7699–7606.

[158] J. Pireaux, J. Riga, R. Caudano, J. Verbist, Electronic Structure of Polymers—X-Ray Photoelectron Valence Band Spectra, in: Photon, Electron and Ion Probes of Polymer Structure and Properties, D.W. Dwight, T.J. Fabish, H.R. Thomas Eds., American Chemical Society Symposium Series, Washington, DC, 1981: pp. 162, 169–202.

[159] J.J. Davis, R. Caudano, X-ray photoemission study of core-electron relaxation energies and valence-band formation of the linear alkanes II Sohd-phase measurements, Phys. Rev. B, 15 (1977) 2242.

[160] J.J. Pireaux, R. Caudano, Experimental picture of the band structure formation in a solid, Am. J. Phys., 52 (1984) 821–826.

[161] G.S. Painter, D.E. Ellis, A.R. Lubinsky, Ab Initio calculation of the electronic structure and optical properties of diamond using the discrete variational method, Phys. Rev. B, 4 (1971) 3610.

[162] G. Calzaferri, R. Rytz, The band structure of diamond, J. Phys. Chem., 100 (n.d.) 1996.

[163] J. Lischner, S. Nemšák, G. Conti, A. Gloskovskii, G.K. Pálsson, C.M. Schneider, W. Drube, S.G. Louie, C.S. Fadley, Accurate determination of the valence band edge in hard X-ray photoemission spectra using GW theory, J. Appl. Phys., 119 (2016) 165703.

[164] E.P.F. Lee, A.W. Potts, M. Dovan, I.H. Hillier, J.J. Delamy, R.W. Hawksworth, M.F. Guest, Photoelectron spectra and electronic structure of the transition metal dichlorides, MCl2(M = Cr, Mn, Fe, Co, Ni), J. Chem. Soc., Faraday Trans. II, 76 (1980) 506.

[165] L.C. Davis, Theory of resonant photoemission spectra of 3d transition-metal oxides and halides, Phys. Rev. B, 25 (1982) 2912.

[166] A.W. Potts, T.A. Williams, W.C. Price, Photoelectron spectra and electronic structure of diatomic alkali halides, Proc. R. Soc. London A, 341 (1974) 147.

[167] C.-Y. Kuo, T. Haupricht, J. Weinen, H. Wu, K.-D. Tsuei, M.W. Haverkort, A. Tanaka, L.H. Tjeng, Challenges from experiment electronic structure of NiO, Eur. Phys. J. Spec. Top., 226 (2017) 2445–2456.

[168] P. Fulde, Electron Correlations in Molecules and Solids, 2nd Edition, Springer, Berlin; Heidelberg, 1993.

[169] V.A. Coleman, C. Jagadish, Chapter 1 – Zinc Oxide Bulk, Thin Films and Nanostructures: Basic Properties and Applications of ZnO, in: Zinc Oxide Bulk, Thin Films and Nanostructures: Processing, Properties and Applications, Elsevier Ltd., Oxford, 2006: pp. 1–20.

[170] R. Rusdi, A.A. Rahman, N.S. Mohamed, N. Kamarudin, N. Kamarulzaman, Preparation and band gap energies of ZnO nanotubes, nanorods and spherical nanostructures, Powder Technol., 210 (2011) 18–22.

[171] N. Kamarulzaman, M.F. Kasim, N.F. Chayed, Elucidation of the highest valence band and lowest conduction band shifts using XPS for ZnO and Zn099Cu001O band gap changes, Res. Phys., 6 (2016) 217–230.

[172] N. Kamarulzaman, M.F. Ksim, R. Rusdi, Band gap narrowing and widening of ZnO nanostructures and doped materials, Nanoscale Res. Lett., 10 (2015) 346.

[173] F.K. Shan, G.X. Liu, W.J. Lee, B.C. Shin, Stokes shift, blue shift and red shift of ZnO-based thin films deposited by pulsed-laser deposition, J. Cryst. Growth, 291 (2006) 328–333.

[174] R.G. Egdell, J. Rebane, T.J. Walker, D.S. Law, Competition between initial- and final-state effects in valence- and core-level X-ray photoemission of Sb-doped SnO2, Phys. Rev. B, 59 (1999) 1792.

[175] C. Kittel, Introduction to Solid State Physics, 8th Edition, John Wiley & Sons, Hoboken, NJ, 2005.

[176] M. Alonso, E.J. Finn, Quantum and Statistical Physics, Addison-Wesley Publishing Company, Boston, MA, 1968.

[177] O. Madelung, Introduction to Solid-State Theory, 3rd Edition, Springer, Berlin; Heidelberg; New York, 1996.

[178] N.W. Ashcroft, N.D. Mermin, Solid State Physics, 1st Edition, Saunders College Publishing, Orlando, FL, 1976.

[179] C.S. Fadley, Instrumentation for surface studies: XPS angular distributions, J. Electron Spectrosc. Relat. Phenom., 5 (1974) 725–754.

[180] Wikipedia Free Encyclopedia, Brillouin Zone, https://en.wikipedia.org/wiki/Brillouin_zone, n.d.

[181] M.C. Biesinger, B.P. Payne, A.P. Grosvenor, L.W.M. Lau, A.R. Gerson, R.S.C. Smart, Resolving surface chemical states in XPS analysis of first row transition metals, oxides and hydroxides: Cr, Mn, Fe, Co and Ni, Appl. Surf. Sci., 257 (2011) 2717–2730.

[182] M. Oku, S. Suzuki, N. Ohtsu, T. Shishido, K. Wagatsuma, Comparison of intrinsic zero-energy loss and Shirley-type background corrected profiles of XPS spectra for quantitative surface analysis: Study of Cr, Mn and Fe oxides, Appl. Surf. Sci., 254 (2008) 5141–5148.

[183] N. Weidler, J. Schuch, F. Knaus, P. Stenner, S. Hoch, A. Maljusch, R. Schafer, B. Kaiser, W. Jaegermann, X-ray photoelectron spectroscopic investigation of plasma-enhanced chemical vapor deposited niox, NiOx(OH)y, and CoNiOx(OH)y: Influence of the chemical composition on the catalytic activity for the oxygen evolution reaction, J. Phys. Chem. C, 121 (2017) 6455–6463.

[184] G.A. Vernon, G. Stuky, T.A. Carlson, Comprehensive study of satellite structure in the photoelectron spectra of transition metal compounds, Inorg. Chem., 15 (1976) 278–284.

[185] M.A. Brisk, A.D. Baker, Shake-up satellites in X-ray photoelectron spectroscopy, J. Electron Spectrosc. Relat. Phenom., 7 (1975) 197–213.

[186] A.P. Grosvenor, R.G. Cavell, A. Mar, X-ray photoelectron spectroscopy study of the skutterudites LaFe4Sb12, CeFe4Sb12, CoSb3, and CoP3, Phys. Rev. B, 74 (2006) 125102.

[187] B. Johansson, N. Martensson, Nature of the 3d shake-down satellites in the lanthanide metals comments on the use of photoelectron spectroscopy for analysis of mixed-valence behavior, Phys. Rev. B, 24 (1981) 4484.

[188] F. de Groot, A. Kotani, Core Level Spectroscopy of Solids, RC Press Taylor & Francis Group, Boca Raton, FL; London; New York, 2008.

[189] C. Godet, D.G.F. David, V.M. da Silva Santana, J. Souza de Almeida, D. Billeau, Photoelectron Energy Loss Spectroscopy: A Versatile Tool for Material Science, in: Recent Advances in Thin Films, Sushil Kumar, D.K. Aswal Eds., Springer Nature, Singapore, 2020.

[190] A. Schleife, C. Rodl, F. Fuchs, J. Furthmuller, F. Bechstedt, Optical and energy loss spectra of MgO, ZnO, and CdO from ab initio many-body calculations, Phys. Rev. B, 80 (2009) 035112.

[191] D. David, C. Godet, Derivation of dielectric function and inelastic mean free path from photoelectron energy-loss spectra of amorphous carbon surfaces, Appl. Surf. Sci., 387 (2016) 1125–1139.

[192] L. Calliari, S. Fanchenko, M. Filippi, Plasmon features in electron energy loss spectra from carbon materials, Carbon, 45 (2007) 1410–1418.

[193] J.A. Leiro, M.H. Heinonen, T. Laiho, I.G. Batirev, Core-level XPS spectra of fullerene, highly oriented pyrolitic graphite, and glassy carbon, J. Electron Spectrosc. Relat. Phenom., 128 (2003) 205–213.

[194] H.R. Philipp, E.A. Taft, Optical properties of diamond in the vacuum ultraviolet, Phys. Rev. B, 127 (1962) 159–161.

[195] D.G.F. David, M.A. Pinault-Thaury, D. Ballutaud, C. Godet, Sensitivity of photoelectron energy loss spectroscopy to surface reconstruction of microcrystalline diamond films, Appl. Surf. Sci., 273 (2013) 607–612.

4 Quantification and Reporting

4.1. QUANTIFICATION

As for other characterization techniques, it is desirable to relate the measured signal intensities to the quantity of various species composing the analyzed material. Besides to the bond assignment, the estimation of the abundance of the chemical species in a material is one of the strength of the XPS technique. In XPS, dealing with quantification, *sensitivity factors* must be used to compare signals from different elements. In this analytical technique, the collected signal depends on a number of parameters related to the nature of the emitting species and the non-isotropic photoelectron emission. The photoelectron current measured by the spectrometer generated by incident X-ray photons of energy, hv, ionizing a core level x of element A in a solid can be expressed by the relation:

$$I_A(x) = I_0 p\sigma(hv, E_x) L(hv, x) \Delta\Omega\, T(E_x)\, D(E_x)\int_0^\infty N_A(z)$$

$$\exp[-z/\lambda(E_x)\cos\theta]dz \qquad\qquad 4.1$$

where I_0 is the X-ray flux, illuminating the sample, p is a parameter accounting for the surface roughness (shadowing effects influence both the X-ray illumination and the photoelectron ejection), $\sigma(hv, E_x)$ is the photoionization cross section of orbital x by photons hv, and $L(hv, x)$ is the associated angular asymmetry factor for emission from x. $\Delta\Omega$ is the acceptance angle of the analyzer and $T(E_x)$ is the spectrometer transmission efficiency for analyzed area of the sample while $D(E_x)$ is the detector efficiency at kinetic energy E_x. $N_A(z)$ is the distribution of element A with the depth z; $\lambda(E_x)$ is the inelastic mean-free path (IMFP) of electrons at energy E_x emitted in the solid containing the atom A. Finally, θ is the emission angle defined by the photoelectron direction and the vertical to the sample surface.

It is reasonable to assume that the sample has a homogeneous composition, and the X-photon flux is constant. Then the integral over z gives simply $N_A\lambda(E_x)\cos\theta$ and the signal intensity measured by the analyzer can be simplified to

$$I_A = [p\,\sigma\, L\, N_A\,\lambda(E_x)\,\cos\theta]\cdot[I_0\Delta\Omega\, T(E_x)\, D(E_x)] \qquad 4.2$$

The first part of these two contributions is sample dependent, while the other is instrument dependent. From equation (4.2), it clearly appears that the intensity of the photoelectric peak I_A is proportional to the number of emitting atoms N_A. However, it is not easy to obtain the value of N_A measuring the intensity I_A because it is not possible to evaluate the parameter p. An easy solution to overcome this problem is to refer the intensity of a generic element to that of a reference element R. I_R is the

DOI: 10.1201/9781003296973-4

intensity obtained from a pure specimen with a clean, smooth surface, just composed by element R. In this case, we can measure the ratio

$$I_A / I_R = \left[p\sigma_A L_A N_A \lambda(E_A)\cos\theta\, I_0 \Delta\Omega T_A(E_A)D(E_A) \right] /$$

$$\left[p\sigma_R L_R N_R \lambda(E_R)\cos\theta\, I_0 \Delta\Omega T(E_R)D(E_R) \right] \qquad 4.3$$

If we measure the intensity ratio in homogeneous pure A and R samples, the parameters p and $\Delta\Omega$ disappear as well as $cos\theta$ and I_0. Then, equation (4.3) simplifies to

$$I_A / I_R = N_A / N_R \cdot [\sigma_A L_A \lambda(E_A)\, T(E_A)D(E_A)] / [\sigma_R L_R \lambda(E_R)T(E_R)D(E_R)] \quad 4.4$$

If we assign a well-defined value for the I_R reference intensity, we then may correlate the intensity I_A of the photopeak A from the orbital x to the ratio N_A/N_R. Generally, F 1s is taken as a reference element and by definition $I(F\ 1s)$ is fixed to 1. The ratio

$$S_A(x) = \left[\sigma_A L_A \lambda_A T(E_A)D(E_A) \right] / \left[\sigma_F L_F \lambda_F T(E_F)D(E_F) \right] \qquad 4.5$$

defines the *sensitivity factor* of element A with respect to F 1s [1, 2]. Then I_A/I_R depends only on the ratio N_A/N_R defining the concentration of element A with respect to N_R which for a pure reference specimen is equal to 1 (100% of R). As an example, in a pure compound with known stoichiometry as the polytetrafluoroethylene—(CF_2-CF_2)—the *sensitivity factor* of carbon can be determined relatively to fluorine as the intensity ratio I_C/I_F corrected by a factor of 2: $S_C = 2 \cdot I_C / I_F$. The same procedure can be applied to all the elements as done by Wagner [3]. It is then possible to correlate the intensity I_A to N_A through S_A

$$I_A = S_A(x)N_A \qquad 4.6$$

S_A is the *sensitivity factor* of the orbital x of the element A. Modern instruments are also provided with the *transfer function* of the analyzer making possible to correct the spectral intensities for the dependence on the analyzer response. The transfer function of the analyzer describes the efficiency of the instrument in collecting photoelectrons at a given kinetic energy. This efficiency depends on how the electrostatic lenses drive and focus the photoelectrons to the entrance slit of the analyzer. Also, electrostatic lenses accelerate/decelerate photoelectrons to a specific energy, the pass energy, so that the error done by the analyzer in measuring the kinetic energy is the same over the acquisition range. Manufacturers defined appropriate functions to tune the voltages of the electrostatic lenses to obtain the best instrument efficiency. However, the efficiency changes upon selection of different instrument operating modes. The analyzer transfer function provided for each of the lens modes and energy resolution (pass energies) allows comparing spectra obtained in different operative conditions. Concerning the detector, its behavior is almost constant with the energy. So far, the experimental *sensitivity factors* will account for the difference in the atom's cross section corrected for the asymmetry factor L, and the different IMFP and the analyzer response. Besides the experimental evaluation, the values of S_i are also determined theoretically as done by Scofield [4], although their value leads to less precise evaluation of the element concentrations. A complete table of S_i for all the elements is commonly provided by the instrument manufacturer since they are specific to the characteristics of the photoelectron collection system, on the spectrometer and detector response. Known the *sensitivity factors*, the concentration of the generic element A is described by the relation

$$X_A = \frac{I_A/S_A}{\sum_i I_i/S_i} \qquad 4.7$$

where I_A and I_i are the intensities of elements A, and with i running on all the chemical species detected in the analyzed specimen. In equation (4.7), the number of emitting atoms per unit volume is determined by dividing the correspondent signal intensity by the relative sensitivity factor. Then the corrected signal can be compared to the total of the corrected signals measured, and a composition is determined. Values obtained by this method are referred to as *atomic concentrations*.

The relation (4.6) requires the intensities of all elements to be carefully measured. In addition to the primary spectrum resulting from non-scattered or elastically scattered electrons, there are intrinsic excitations which take place during the propagation of the electrons toward the surface. Then, an energy transfer of the core-photoelectrons via intrinsic and extrinsic excitation processes with generation of the loss-features as described in Section 3.6. As a consequence, the spectrum is formed by a principal peak followed by a tail at lower binding energies. Extrinsic losses are due to inelastic-scattering and secondary electron cascade processes which contribute to a background extending to the whole wide spectrum. Selection of an appropriate baseline for background subtraction is then essential for a correct evaluation of the area under a generic peak and then for the estimation of the element concentrations.

4.2.   FACTORS AFFECTING THE QUANTIFICATION

The power of XPS is the possibility of providing a list of information on the elemental composition of the surface of a specimen. However, a correct evaluation of the composition requires compliance with some basic rules:

(a) the need to correctly prepare the sample to avoid contaminations which could hinder the real composition of the material;
(b) optimal acquisition conditions to generate spectra with acceptable signal-to-noise ratio;
(c) evaluation if the sampling depth and the analyzed volume is sufficient to describe the characteristics of the entire sample;
(d) sample damaging due to high-energy irradiation with X-photons.

Sample cleaning is generally applied using solvents to remove organic contaminations when possible. Also the selection of a good solvent may be important since residuals from evaporation likely remain on the sample surface. To have an idea about the influence of solvent residuals on the XPS spectra, a good idea is to select a solvent containing an element that can be used as a marker to quantify residuals. As an example, if Cl is not part of the specimen chloroform or trichloroethylene can be a solution. In the case of organic compounds situation is more complex because not always is possible to apply surface cleaning procedure without damaging the sample. In these cases, the only solution is to prepare the samples and store them under an inert gas as argon or nitrogen in a clean box until they are inserted in the instrument under vacuum. Minimization of the exposure to atmosphere is always a good rule, because this minimizes the hydrocarbon contamination. Selection of the optimal acquisition conditions is related to optimization of X-irradiation with elimination of possible shadows in corrugated samples. This is done by a correct orientation of the sample toward the X-source. Influence of the surface roughness is important especially at high sample inclination angles when angle-resolved XPS is performed. When the take-off angle is small, possible interferences of the sample roughness with the photoelectron path toward the analyzer are possible.

In Section 3.5.3, we already discussed the mean-free path of electrons in its propagation in a solid matrix. This is strictly related to the material density and led Sea M. and Dench W.A. to define a *universal curve* to describe the inelastic mean-free path for different classes of solids [5, 6]. It is then possible to estimate the attenuation length and the related sampling depth. If, in the analyzed specimen, the element abundance is not homogeneous with the depth, the quantification cannot represent the real composition. An important remark regards also the need of a sufficient high SNR which may require long acquisition times. X-radiation is sufficiently strong to break chemical bonds inducing sample damaging. This problem becomes important especially in organic compounds where visible degradations of the sample surface can appear in the irradiated area. If the sample is homogeneous, it is possible to change the sample position to irradiate a new fresh zone thus providing a solution for the problem of degradation.

4.3. SAMPLING DEPTH AND ATTENUATION LENGTH

The quantification process is based on the contribution of the different chemical elements of the material. As described in Section 1.3.2, the different kinetic energies of the photoelectrons result in different sampled volumes which should be taken into account in a correct quantification process. The inelastic mean-free path (IMFP) described in Section 1.3.2 ignores elastic scattering. However, the description given by the *Universal Curve* [5] expressed by equation (1.11) and all the next attempts to interpret the ensemble of loss processes on the basis of experimental data [7–10] include the elastic scattering resulting in the *Attenuation Length* or better the *Effective Attenuation Length* (EAL). In addition, the application of the Beer law to the decay of the photoelectric signal (see equation 1.8 in Section 1.3.2) is based on the assumption that the emission and the propagation of the photoelectron in the solid is isotropic. In reality, in XPS the photoionization process is anisotropic and the effects of elastic scattering are particularly prominent. Inelastic and elastic scattering are co-existent processes which have to be considered together to describe the EAL λ. Form one side, the presence of electron elastic scattering increases the total path length before emission from the solid. On the other hand, increasing the path traveled also increases the inelastic scattering probability. Then to get the EAL needed for quantitative AES and XPS, a separate term accounting for the elastic scattering should be considered when starting from IMFP values. An estimation of the EAL can be obtained from the IMFP Λ by the relation [11]

$$\lambda = \Lambda\left(1 - 0.028\sqrt{Z}\right)\left[0.501 + 0.068 \ln\left(E_K\right)\right] \qquad 4.8$$

Equation (4.8) shows that the EAL λ decreases with the atomic number Z while it increases with the kinetic energy E_K. Still the relation (4.8) is a crude approximation of the EAL, because it does not consider the dependence on the emission angle θ. The exact definitions of the relevant terms are described in [12–14]. Also, equation (4.8) still assumes an exponential decay of the intensity with the depth. In contrast to inelastic scattering, the elastic scattering contribution has a non-exponential dependence on the depth and on the emission angle θ. Therefore, it has become customary to treat both parts separately and to describe the attenuation length λ by a product of the IMFP Λ and an "elastic scattering correction factor" Q [15, 16]. However, due to anisotropic behavior of the photoelectron emission in XPS, the elastic scattering correction factor must be multiplied by an additional term to account for the effect of elastic scattering on the asymmetry of photoelectron emission. The parameter Q depends on the emission angle θ and on the material and is described by the relation

$$\lambda = \Lambda\, Q\left(\theta,\, \omega\right) \qquad 4.9$$

with

$$Q\left(\theta,\, \omega\right) = \left(1 - \omega\right)^{1/2} H\left(\cos\theta,\, \omega\right) \qquad 4.10$$

H(cosθ, ω) is the Chandrasekar function

$$H\left(\cos\theta,\, \omega\right) = \left(1 + 1.9078\right)/\left[1 + 1.9078\cos\theta\left(1 - \omega\right)^{1/2}\right] \qquad 4.11$$

ω is the single-scattering albedo defined by

$$\omega = 1/\left[1 + \Lambda_{TR}/\Lambda\right] \qquad 4.12$$

with the inelastic mean-free path Λ and the transport mean-free path Λ_{TR}. For the measurement of overlayer-film thickness by XPS, the practical EAL do not vary significantly with the overlayer-film thickness for emission angles in the range $0 < \theta < 50^0$ [17]. It is therefore convenient computing average values of EALs that can be applied in XPS or AR-XPS experiments for useful values of overlayer thicknesses over the same range of emission angles. The values of inelastic mean-free paths Λ (nm) and transport mean-free paths Λ_{TR} (nm) can be estimated using the theory of electron transport in solids [18]. Authors calculated the EAL for four

TABLE 4.1

Kinetic Energies, Inelastic Mean-Free Paths Λ(nm), Transport Mean-Free Paths Λ_{TR}(nm), Values of the Single-Scattering Albedo, ω, and Values of the Asymmetry Parameter, β, for the Photoelectron Lines Considered in the Cited Work

Core-line	E_K(eV)	Λ(nm)	Λ_{TR}(nm)	ω	β
Mg Kα X-rays					
Si $2s_{1/2}$	1104	2.663	17.336	0.1332	1.997
Si $2p_{3/2}$	1154	2.758	18.528	0.1296	1.085
Cu $2s_{1/2}$	157	0.565	0.717	0.4404	1.990
Cu $2p_{3/2}$	321	0.795	1.045	0.4320	1.364
Cu $3p_{3/2}$	1179	1.912	4.485	0.2988	1.561
Ag $3s_{1/2}$	535	0.920	1.714	0.3492	1.979
Ag $3p_{3/2}$	681	1.081	2.152	0.3344	1.553
Ag $3d_{5/2}$	885	1.300	2.830	0.3148	1.198
Ag $4s_{1/2}$	1157	1.582	3.837	0.2919	1.985
Au $4s_{1/2}$	491	0.825	1.710	0.3255	1.811
Au $4p_{3/2}$	707	1.053	1.951	0.3505	1.253
Au $4d_{5/2}$	918	1.266	2.294	0.3556	1.131
Au $4f_{7/2}$	1170	1.511	2.777	0.3524	1.004
Al Kα X-rays					
Si $2s_{1/2}$	1337	3.099	23.160	0.1180	1.995
Si $2p_{3/2}$	1387	3.191	24.498	0.1153	1.008
Cu $2s_{1/2}$	390	0.892	1.239	0.4186	1.994
Cu $2p_{3/2}$	554	1.118	1.776	0.3864	1.445
Cu $3p_{3/2}$	1412	2.189	5.720	0.2768	1.526
Ag $3s_{1/2}$	768	1.176	2.432	0.3258	1.981
Ag $3d_{5/2}$	914	1.331	2.932	0.3122	1.607
Ag $3d_{5/2}$	1118	1.542	3.686	0.2949	1.209
Ag $4s_{1/2}$	1390	1.815	4.783	0.2751	1.986
Au $4s_{1/2}$	724	1.070	1.976	0.3514	1.829
Au $4p_{3/2}$	940	1.288	2.333	0.3556	1.389
Au $4d_{5/2}$	1152	1.494	2.740	0.3528	1.224
Au $4f_{7/2}$	1403	1.731	3.273	0.3459	1.030

Source: Reprinted with permission from [18]

elemental solids (Si, Cu, Ag, and Au) and for energies between 160 eV and 1.4 keV and considering the XPS anisotropy. The results are summarized in Table 4.1.

For overlayer-thickness determination, EALs were calculated applying Monte Carlo models [17]. Results show that the EALs were almost independent of the emission angle up to $\theta = 50°$ while it sharply increases for $\theta > 50°$. Finally for $0° \leq \theta \leq 50°$, the EAL depends very weakly on the overlayer thickness. Due to the weak variation of EAL on the emission angle, it is possible to compute an average value for the EAL, $\langle\lambda\rangle$. For the overlayer, the authors computed the ratio

$$R_{TH} = \langle\lambda_{TH}\rangle / \Lambda \qquad\qquad 4.13$$

which can be fitted as a function of the single-scattering albedo ω

$$R_{TH} = 1 - A_{TH}\omega \qquad\qquad 4.14$$

where ω is a convenient measure of the strength of elastic-scattering effects, and calculated the ratio R_{QA}

$$R_{QA} = \langle\lambda\rangle / \Lambda \qquad\qquad 4.15$$

which can be fitted by a second order function of ω

$$R_{QA} = 1 - A_{QA}\omega - B_{QA}\omega^2 \qquad\qquad 4.16$$

Results are shown in Figure 4.1. The values EAL $\langle\lambda\rangle$ for quantitative determination of surface composition by AES and XPS for other materials can be simply found from these relationships.

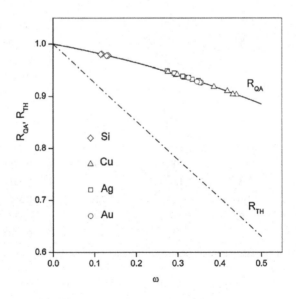

FIGURE 4.1 Comparison of predictive formulas for EALs for quantitative analysis by AES and XPS as described in the text and for overlayer-thickness determination. Symbols: ratios R_{QA} obtained from averaging the EALs for quantitative AES and XPS according to equation (4.11); solid line: fit of equation (4.12) to the R_{QA} values; dotted-dashed line; predictive formula for EAL for measurements of an overlayer thickness according to (4.10) with $A_{TH} = 0.738$ [17].

Equation (4.14) allows obtaining λ as a function of the IMFP as $\lambda = \Lambda \left(1 - A_{TH}\omega\right)$. The term $\left(1 - A_{TH}\omega\right)$ represents the elastic scattering correction factor and A_{TH} was estimated to be a constant equal to 0.713 [17]. Generally for AES and XPS measurements, ω is between 0.05 and 0.45 and then the elastic scattering correction term $(1 - A_{TH}\omega)$ ranges between 0.96 and 0.68. From these relations, it is possible to obtain a mean electron escape depth for isotropic emission expressed by

$$\lambda_{MED} = \Lambda \left(1 - A_{TH}\omega\right) \cos\theta \qquad (4.17)$$

λ_{MED} is the value of the effective attenuation length which should be used in equation (4.1) to describe the signal intensity obtained by a photoemission process.

4.4. TARGET FACTOR ANALYSIS

Chemometrics is the application of mathematical and statistical methods to the analysis of chemical data. Among the statistical approaches, the target factor analysis (TFA) has proven to be one of the powerful methodologies to interpret and analyze spectral data. Essentially the TFA is a multivariate technique to reduce matrices of data to their lower dimensionality by the use of orthogonal factors and transformation providing those factors [19]. In electron and photoelectron spectroscopy, the TFA become popular after the pioneering work of Gaarenstroom [20] because of the ability to decompose large data sets of complicated spectra into meaningful spectral components [20–23]. TFA together with principal component analysis, spectral synthesis, and linear least squares fitting demonstrated their usefulness in quantitative analysis for typical areas such as chemical mapping, depth profiling, oxidation, implantation, and background subtraction. In particular, TFA is used to extract the number and the quantity of characteristic spectral features contained in a matrix of data. As a consequence, its main applications are the recognition of chemical bonding features in AES and XPS array of spectra [23]. In the following, we will use the formalism described by Malinowski and Howery for chemometrics [19].

As anticipated in the previous paragraph, the scope of TFA is the determination of the significant factors (components) and their respective concentrations (loadings). The loadings turn out to be the coefficients of the linearly combination which can be used to reproduce the whole sequence of measured spectra as a function of an outer parameter. This latter may represent a synthesis parameter like the extent of oxidation or sputtering time or the annealing temperature. The array of M spectra, each composed by N data, constitutes the columns of the matrix $[D]$ with dimension (N × M). Application of TFA will decompose $[D]$ in a product of two matrices

$$[D] = [R] \times [C] \qquad (4.18)$$

In equation (4.18), $[R]$ represents the matrix of the relevant components, while $[C]$ is the matrix of the weighting factors. The decomposition of $[D]$ is made using the principal component analysis (PCA) [19]. Using appropriate mathematical procedures is possible to determinate the minimum number of "eigenvectors" *C* to reproduce the measured spectra. The result of the TFA applied to a series of measured spectra is (i) the number of independent component spectra and (ii) the synthesis of the measured spectra obtained by a linear composition of the principal components. Essentially TFA computes the covariance matrix [Z] obtained multiplying [D] for its transpose [D]^t. The number of primary factors is found taking those with higher eigenvalues. There is a list of tests to determine the number of relevant factors. Among others, the *Imbedded Error* (IE) and the *Indicator function* (IND), *Percentage Variance* (PV), and *Cumulative Percentage Variance* (CPV) methods recognize the information carried by spectra above a noise limit [19, 24]. The correspondent eigenvectors are the scaled projection of the base independent spectra on the measured spectra. Scaling factors correspond to the peak area of the components normalized by the correspondent sensitivity factors, thus resulting in the element concentrations. Once the reduced eigenvector matrix is determined, the matrix [*D*] is obtained as

$$[D] = [R'] \times [C']$$

(4.19)

Observe that here [*R'*] represents an abstract matrix of data namely *non-physical spectra*. A target transformation (TT) is then applied to the abstract matrix to obtain real spectra [25]. Then resolving the [*C'*] concentration matrix, TT will provide the real concentrations. The TFA is non-implemented in the free version of the *RxpsG* software. Authors interested in this kind of analysis are invited to contact the author. Also, an alternative solution is the use of the R specific package to perform the Factor Analysis.

4.5. SIMPLE QUANTIFICATION PROCESS, QUANTIFICATION REPORT

Quantification is found among the options of the *Analysis* menu and the correspondent GUI is shown in Figure 4.2. As usual on the top of the GUI, the XPS Sample to make the quantification can be selected.

As it can be observed, the GUI is formed by a sequence of pages, each corresponding to the elements where a background subtraction was performed. In each page on the left is indicated the core-line name checked by default. Figure 4.2 shows the notebook page corresponding to the carbon core-line. In the frame components are indicated all the components used for fitting the core-line. All the components are also considered by default in the quantification, but the user can disregard one or more components just un-checking them. Finally for each of the fit components, is indicated the *relative sensitivity factor* (RSF) (Figure 4.2a) which will be used for the quantification. The default RSF corresponds to the values provided by the instrument manufacturer and can be freely modified when necessary. Observe that if for a given element just the baseline was added and no fit performed then in the GUI the *Components* frame will be empty and just one RSF will be used for the whole peak as illustrated in Figure 4.2b for N 1s. The last page is dedicated to the quantification. It may happen that, when analyzing the data, the user realizes that unexpected or

(a)

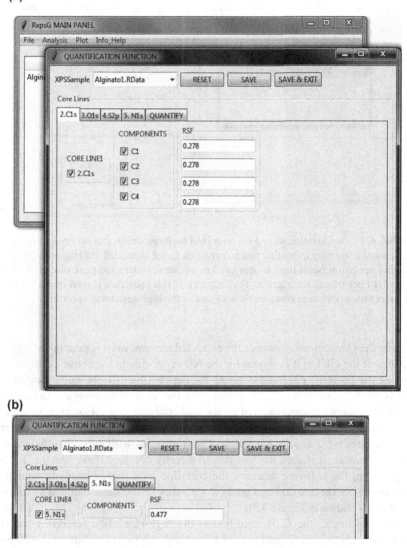

(b)

FIGURE 4.2   (a) The graphical interface of the *Quantification* option relative to the C1s element. (b) In the case of the N 1s element no fit component are present since just the baseline is defined. Only the *Relative Sensitivity Factor* can be modified.

less-important elements characterized by very small intensities were not acquired in high resolution. It is possible to obtain a crude estimation of the relative abundance by *extracting* the correspondent peak from the survey as illustrated in Section 3.4.3. In this case the energy resolution is scarce and generally, although possible, no fit is performed. To include the extracted peak in the quantification, a baseline must be defined as commonly done for high-resolution spectra (see Section 3.7.2). Then

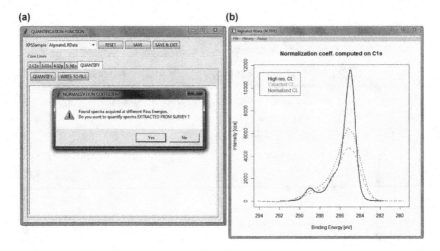

FIGURE 4.3   (a) Quantification done on mixed high-resolution and survey-extracted core-lines caused a warning indicating that a correction factor is needed. (b) High-resolution C 1s reference spectrum (solid line) is compared to the survey-extracted peak of carbon (dashed line) after background subtraction. The intensity of this spectrum is then corrected (dotted line) to obtain a peak area comparable with that of the high-resolution spectrum.

opening the *Quantification* option, the extracted element must appear in the sequence of pages of the GUI as it is shown for the N 1s core-line in the above example. Now opening the *Quantify* page and pressing the button *Quantify*, the quantification process is started. For core-lines acquired with the same pass-energies, the software readily computes the abundance of all the analyzed core-lines. However, in presence of survey-extracted peaks, a warning is lit (Figure 4.3a). The software recognizes that core-lines were acquired at different pass-energies with different efficiencies of the spectrometer. It is then not possible to directly compare these spectra. To solve the problem, the software searches the core-line with higher intensity among the high-resolution spectra and compares it with that of the same element in the wide spectrum as shown in Figure 4.3b.

In this example, the C 1s core-line is the high-resolution reference spectrum (solid line) which is compared to the background-subtracted carbon peak of the survey (dashed line). The ratio of their intensities gives the normalization factor which will be used to correct the survey-extracted spectrum. In Figure 4.3b, after normalization of the intensity, the survey-extracted and corrected C1s peak (dotted line) should have an area comparable with that of the original high-resolution core-line. Obviously the two spectra cannot overlap because the different energy resolution affecting their FWHM. Now, known the normalization factor, core-lines acquired using different pass-energies can be compared and element quantification can be performed. The process ends with a summary of the peak spectral properties (area, FWHM, the RSF applied, the energy position, and the concentration for the core-line and for each of the fit components) for all the analyzed core-lines as illustrated in Figure 4.4.

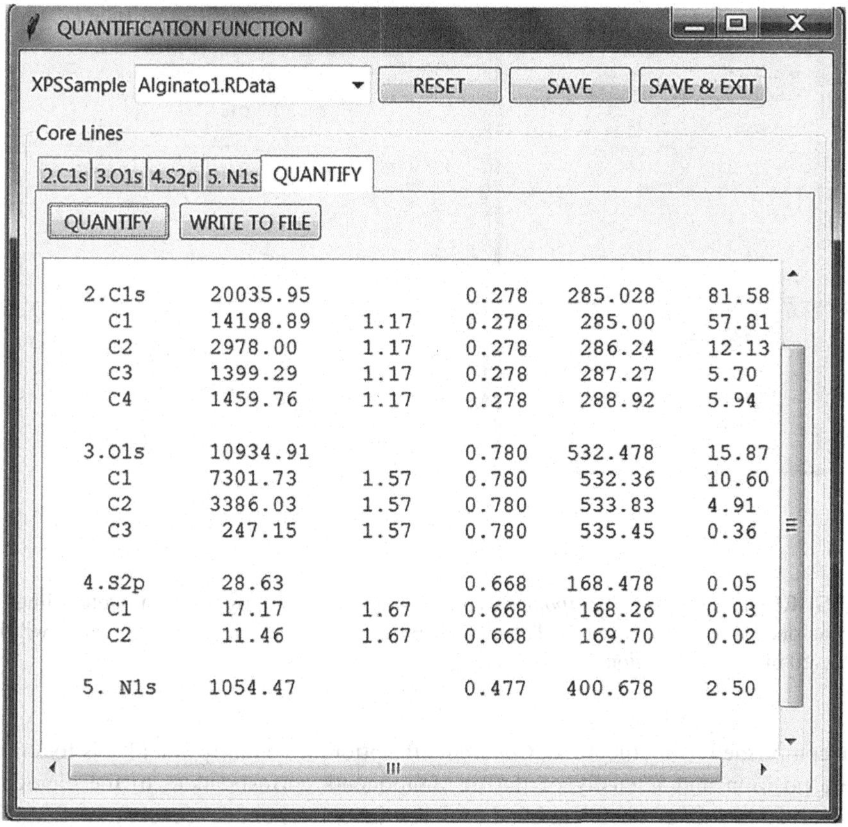

FIGURE 4.4 The quantification GUI showing the quantification report.

The quantification report can be selected, copied, and directly pasted in a document. Observe that the quantification report in the GUI is written using the *Courier monospace* font to obtain aligned data. The same font should be used in the document to align data in columns. In addition to the *Quantify* button, there is a *Write to file* button to export the data in a file if needed.

4.6. MOVE COMPONENTS OPTION AND STOICHIOMETRY

The elemental quantification provides the abundance of the chemical species in the analyzed materials. If the high-energy resolution spectra are fitted, the *Quantification Report* gives the atomic concentration of each fit component, that is, the percentage of atoms involved in the correspondent assigned bond. Frequently the interpretation of the surface chemistry and the bond assignment is a complicated task. This happens in particular when different bonds fall at similar binding energies and the spectrometer is unable to resolve the spectral components. In these cases, the suggested approach is to hypothesize a certain surface composition by fitting the core-line spectra and then check the stoichiometry. This task can be

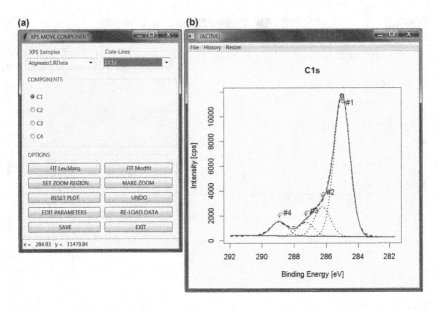

FIGURE 4.5 (a) *Move Components* GUI. (b) Dashed line: original data. Dotted line: fit components. Solid line: best fit. The marker indicates the selected component which will be moved in the new position.

accomplished using the *Move Components* option. Essentially, the idea is to adjust the position and intensity of the fit components correspondent to the elements involved in a given chemical bond to make the correspondent concentrations be in agreement with the expected stoichiometry. As an example, the spectral areas of the carbon and oxygen fit components assigned to the C–O single bond must be in the 1:1 proportion. This process must be repeated for all the different chemical bonds, namely all the fit components of all the chemical species. Figure 4.5a displays the *Move Component* GUI.

On the top, the user can select the XPS Sample and the core-line among the acquired spectra. Upon core-line selection on the left appears the correspondent list of fit components. In the GUI and the graphical window, the active fit component is indicated by a marker (see the component #1 in Figure 4.5b). By left-clicking on the spectrum where to move the active fit component, *RxpsG* will readily place it in the new position. In addition, any change of the fit component position/intensity activates the quantification process

In this manner, the user can slightly modify the fit components and check the effect of these changes on the concentrations. In other words, acting on the fit components of the elements involved in the chemical bond, the user can adjust the spectral area and verify if the stoichiometry is respected. The results are displayed in the *RStudio* console in a typical table format used by the quantification procedure as shown in Figure 4.6. The example of Figure 4.5 shows a C1s fit where the components are assigned to the various carbon–oxygen bonds. The user can start adjusting the C1s fit components and then proceed with those of the oxygen core-line. Referring to

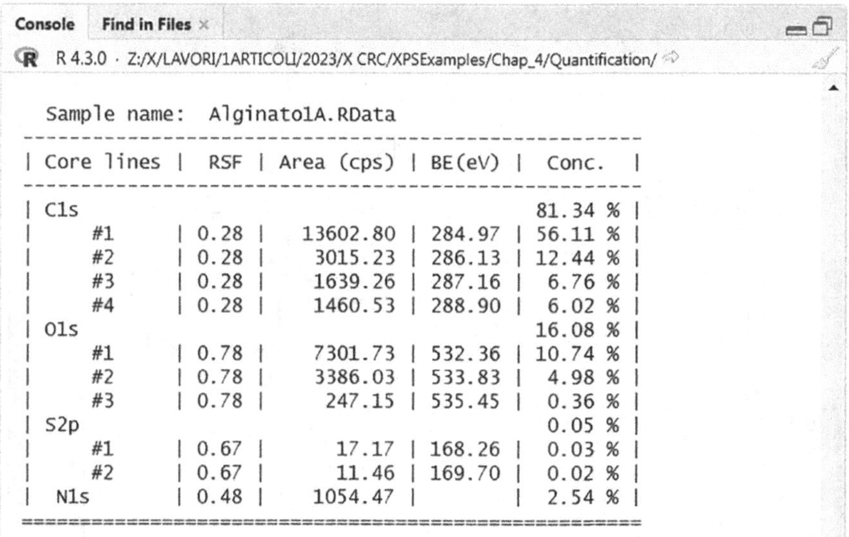

```
Console   Find in Files ×
R 4.3.0 · Z:/X/LAVORI/1ARTICOLI/2023/X CRC/XPSExamples/Chap_4/Quantification/

  Sample name:  Alginato1A.RData
---------------------------------------------------------------
| Core lines |  RSF | Area (cps) | BE(eV) |  Conc.   |
---------------------------------------------------------------
| C1s                                         81.34 % |
|      #1     | 0.28 |   13602.80 | 284.97 | 56.11 % |
|      #2     | 0.28 |    3015.23 | 286.13 | 12.44 % |
|      #3     | 0.28 |    1639.26 | 287.16 |  6.76 % |
|      #4     | 0.28 |    1460.53 | 288.90 |  6.02 % |
| O1s                                         16.08 % |
|      #1     | 0.78 |    7301.73 | 532.36 | 10.74 % |
|      #2     | 0.78 |    3386.03 | 533.83 |  4.98 % |
|      #3     | 0.78 |     247.15 | 535.45 |  0.36 % |
| S2p                                          0.05 % |
|      #1     | 0.67 |      17.17 | 168.26 |  0.03 % |
|      #2     | 0.67 |      11.46 | 169.70 |  0.02 % |
|  N1s        | 0.48 |    1054.47 |        |  2.54 % |
===============================================================
```

FIGURE 4.6 Quantification table generated when the core-line fit component intensity or position is changed.

the quantification table of Figure 4.6, the component C1 is assigned to the C=C graphitic carbon. C2 at 286.6 eV corresponds to the –C–O– and to the –C–OH bonds with a concentration of ~12.5% and follows the C3 at ~287.2 eV assigned to the –O–C–O– bond with a concentration of 6.7%. Finally, the C4 describes the carboxyl –O–(C=O) with a 5.1% abundance. The oxygen fit component correspondent to the C–O single bond is C1 with a concentration of 10.6 %. This component must account not only for the C 1s component C2 but also for the carbonyl C3. However, the alginate monomer is formed by an equal number of carbon and oxygen atoms. It comes out that there is an excess of carbon atoms bonded to oxygen: likely something is lacking. Looking at the list of elements, nitrogen is present but no bonds are assigned to this element while nitrogen should form bonds with carbon. This new carbon component can be added with the *Analysis* GUI and the result is shown in Figure 4.7.

The user can use the *Move Components* to place the C–N component in the correct position with the correct intensity. Therefore, controlling if the stoichiometry is satisfied, indications about the kind and the number of chemical bond assigned are obtained which can confirm or reject the initial hypothesis on the surface chemistry of the material. Before fitting the C 1s, it can be useful to set some constraints to ensure that the new component falls at the correct binding energy, and its FWHM is the same as that of the other components. A first easy solution is to edit the fit parameters of the added component. To do this, the use must select first the desired fit component. Then just pressing the button *Edit Parameters*, the correspondent fit parameters will appear in an editable window as reported in Figure 4.8.

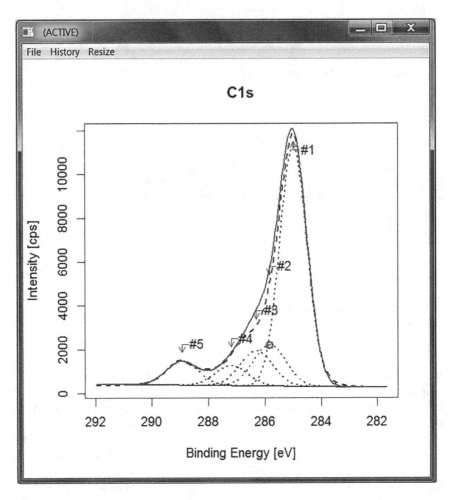

FIGURE 4.7   The C1s core-line with C-N bond component indicated by the marker. The solid line indicates the envelop of the fitting components.

Each of the parameters can be freely modified. Then press *Save* to set the changes and update the core-line plot. In this example, the width of the component C2 was changed to assume the same value equal to that of the other components. Observe that this procedure cannot be used to set fitting constraints. If this is needed for the best fit, the user must utilize the *Fit Constraints* option.

To do this, first the C 1s spectrum must be saved. Then the user must open the *Fit Constraints* option together with the *Move Component* GUI as shown in Figure 4.9. These are independent processes, and the only way to transfer information from one to the other procedure is to save the data in the *Global Environment* (step 2 in Figure 4.9) and *Reload* them into the *Fit Constraints* local memory (step 3 in Figure 4.9). The desired constraints can be set (step 4 in Figure 4.9) and then spectrum

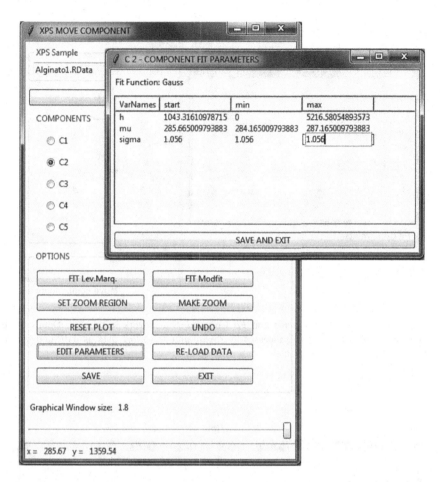

FIGURE 4.8 The *Edit Parameters* button opens an editable window containing the fitting parameters relative to the active fit component.

can be fitted. If further refinements on the stoichiometry are needed, the fitted spectrum must be saved (step 5 in Figure 4.9) and *Reloaded* in the *Move Components* to make the needed changes (step 6 in Figure 4.9).

This procedure can be repeated restarting from (step 1) with optimizing intensity/position of the fitting components. In the present example, the added C-N fit component was refined by placing it at the correct BE, 285.7 eV, and harmonizing the intensity to obtain a concentration in agreement with the N concentration of ~2.5%. After optimization of the C 1s, O 1s, and N 1s components, the results are shown in Figure 4.10. On the left is reported the quantification report. Now the concentration of the C 1s C2 component assigned to C–N–C bonds agrees with the N 1s abundance. Also the sum of the components (C3 + C4) amounts to ~12.5% which agrees quite well with the concentration of the oxygen component C1 ~ 11%. Finally, the

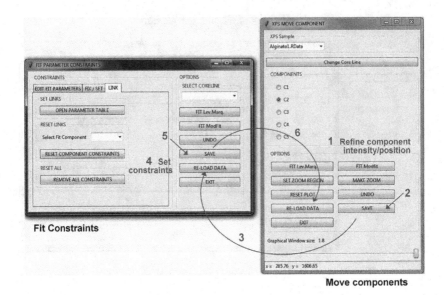

FIGURE 4.9 *Move Components* and *Fit Constraints* can work in parallel. Data can be exchanged between these two processes by *saving* and *reloading* data as indicated by the sequence of numbered operations.

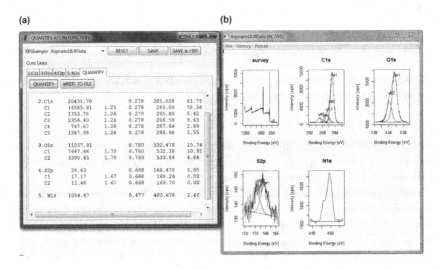

FIGURE 4.10   (a) The quantification results obtained after refining the C 1s fit components. Observe the agreement between the concentration of the carbon C2 and the concentration of N 1s. (b) The XPS Sample core-lines and the spectral analysis accomplished.

carboxyl concentration described by the C1s component C5 ~ 5.4% is in good agreement with the correspondent O 1s component C2 measuring about 5%.

In conclusion, this example shows how the control on the stoichiometry helps to correctly identify the chemical composition of the material. When the concentration of the bond-components is in disagreement with the expected stoichiometry, likely there is an error made by placing the fit components in a wrong position or the identified bonds are wrongly assigned. As shown, a chemical bond was lacking in the C 1s fit to correctly explain the concentration of the oxygen bond components.

4.7. ANALYSIS REPORT

The ability to describe the material is linked to the possibility of using powerful tools to obtain and manipulate the information. In the previous chapters are described the various options available to characterize the material composition and chemistry. *RxpsG* offers also the possibility to compose a report summarizing the results of the spectral analysis. The report is composed by two parts, one summarizes the information regarding the peak fitting and the other displays the results of the quantification process. The *Fit Report* is essentially a table listing the details of the spectral analysis for each of the analyzed core-lines. In the report is indicated the *baseline* applied for the background subtraction and the *fit function* is used for each of the fitting components and the relative parameters, namely position, intensity and area, FWHM, and the relative percentage contribution to the spectral amplitude of the core-line.

The second part of the report contains the quantification results. The quantification is computed considering all the spectral area defined by the *baseline* or, in presence of a best fit, the spectral area corresponding to the background subtracted best fit. The quantification is performed automatically for all the core-lines where a *baseline* was defined. In this option, the user cannot modify the parameters used in the quantification process as the *relative sensitivity factor* or include/exclude fit components, since this must be done using the *Quantification* option. An example of the *Analysis report* is shown in Figure 4.9.

Figure 4.11a shows the graphical user interface of the *Analysis Report*. The software recognizes fitted and non-fitted Core-Lines which appear separated in the two different groups *Fitted* and *Other* (indicated by the circles in the figure). Arrows indicate the correspondent reports that can be obtained. *Fit and Quantification Reports* can be made only for fitted Core-Lines. All the remaining spectra can be described by the *Standard Report*. In Figure 4.11b are shown examples of *Fit Report, Quantification Report*, and *Standard Report*. The *Fit Report* summarizes the list of components used for fitting the spectrum, the correspondent fitting lineshapes, and the fit parameters (component area, intensity, FWHM, and position). Finally, the relative weight is also reported. Assumed the total integral area of the background-subtracted single Core-Line is 100%, the relative weight of each component represents the ratio *component_area/total_area* in percentage. The *Quantification Report* summarizes the quantification result reporting the atomic concentrations of the elements represented by the background-subtracted core-lines. As usual, for fitted core-lines, the report shows the area of each of the fitting components, their FWHM, the relative

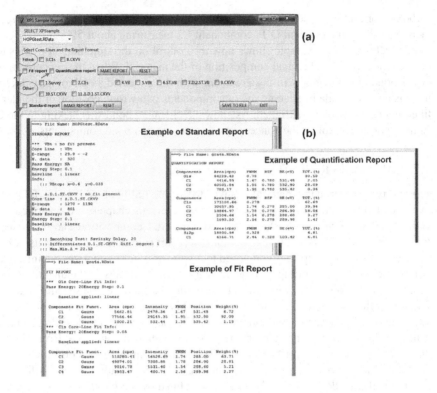

FIGURE 4.11 (a) The *Report* GUI; circles show the list of *Fitted* Core-Line where *Fit Report* and *Quantification Report* can be made, and *Other* non-fitted core-lines where only the *Standard Report* can be obtained. Arrows indicate the checkboxes to obtain the *Fit, Quantification*, and *Standard Reports*.

sensitivity factor applied, the energy position, and the correspondent abundance in atomic percentage. Finally for all the spectra including the non-fitted Core-Lines, the *Standard Report* can be obtained (Figure 4.11b). In this report for each Core-Line are summarized the acquisition conditions namely the acquisition spectral energy range, the energy step, and the pass energy. The *Standard Report* includes also information relative to particular analyses performed as the estimation of the *VB top* or the position of the *VB Fermi Edge*. It describes also, if a spectral smoothing was applied, the kind of filter used and the degree of noise rejection or differentiated spectra, the degree of differentiation, and the position of the derivative maximum/minimum if it was estimated.

REFERENCES

[1] D. Briggs, M.P. Seah, Practical Surface Analysis by Auger and Photoelecron Spectroscopies, John Wiley & Sons, Chichester, 1990.

[2]   D. Briggs, Surface Analysis of Polymers by XPS and Static SIMS, D.R. Clarke, S. Suresh, I.M. Ward Eds., Cambridge University Press, Cambridge, 1998.

[3]   C.D. Wagner, L.E. Davis, M.V. Zeller, J.A. Taylor, R.H. Raymond, L.H. Gale, Empirical atomic sensitivity factors for quantitative analysis by electron spectroscopy for chemical analysis, Surf. Interface Anal., 3 (1981) 211–225.

[4]   J.H. Scofield, Theoretical Photoionization Cross Sections from 1 to 1500 keV, Technical Report UCRL-51326, California University, Lawrence Livermore Laboratory, Livermore, 1973.

[5]   M.P. Seah, W.A. Dench, Quantitative electron spectroscopy of surfaces: A standard data base for electron inelastic mean free paths in solids, Surf. Interface Anal., 1 (1979) 2–11.

[6]   M.P. Seah, An accurate and simple universal curve for the energy-dependent electron inelastic mean free path, Surf. Interface Anal., 44 (2012) 497–503.

[7]   S. Tanuma, C.J. Powell, D.R. Penn, Calculations of electron inelastic mean free paths V Data for 14 organic compounds over the 50–2000 eV range, Surf. Interface Anal., 21 (1994) 165–176.

[8]   S. Tanuma, C.J. Powel, D.R. Penn, Calculation of electron inelastic mean free paths (IMFPs) VII reliability of the TPP-2M IMFP predictive equation, Surf. Interface Anal., 35 (2003) 268–275.

[9]   C.J. Powell, Practical guide for inelastic mean free paths, effective attenuation lengths, mean escape depths, and information depths in X-ray photoelectron spectroscopy, J. Vac. Sci. Technol. A, 38 (2020) 023209.

[10] H.T. Nguyen-Truong, Penn algorithm including damping for calculating the electron inelastic mean free path, J. Phys. Chem. C, 119 (2015) 7883–7887.

[11] A. Jablonski, Effects of Auger electron elastic scattering in quantitative AES, Surf. Sci., 188 (1987) 164–180.

[12] A. Jablonski, C.J. Powel, The electron attenuation length revisited, Surf. Sci. Rep., 47 (2002) 33.

[13] A. Jablonski, C.J. Powel, Practical expressions for the mean escape depth, the information depth, and the effective attenuation length in Auger-electron spectroscopy and X-ray photoelectron spectroscopy, J. Vac. Sci. Technol. A, 27 (2009) 253.

[14] ISO 18115, Surface Chemical Analysis Vocabulary, ISO-Geneva, 2001.

[15] A. Jablonski, Universal quantification of elastic scattering effects in AES and XPS, Surf. Sci., 364 (1996) 380–395.

[16] A. Jablonski, C.J. Powel, Elastic photoelectron-scattering effects in quantitative X-ray photoelectron spectroscopy, Surf. Sci., 606 (2012) 644–651.

[17] A. Jablonski, C.J. Powel, Effective attenuation lengths for photoelectrons emitted by high-energy laboratory X-ray sources, J. Electron Spectrosc. Relat. Phenom., 199 (2015) 27–37.

[18] A. Jablonski, C.J. Powel, Effective attenuation lengths for quantitative determination of surface composition by AES and XPS, J. Electron Spectrosc. Relat. Phenom., 218 (2017) 1–12.

[19] E.R. Malinowski, D.G. Howery, Factor Analysis in Chemistry, 3rd Edition, Wiley, New York, 2002.

[20] S.W. Gaarenstroom, Quantitative analysis of sputtered a- and a+b-brass surfaces by using Auger electron spectroscopy with principal component analysis-target factor analysis, J. Vac. Sci. Technol. A, 16 (1979) 600.

[21] S. Hofmann, J. Steffen, Factor analysis and superposition of Auger electron spectra applied to room temperature oxidation of Ni and NiCr21Fe12, Surf. Interface Anal., 14 (1989) 59–65.

[22]   C. Palacio, H.J. Mathieu, Application of factor analysis to the AES and XPS study of the oxidation of chromium, Surf. Interface Anal., 16 (1990) 178–182.

[23]   W.F. Stickle, The Use of Chemometrics in AES and XPS Data Treatment, in: Surface Analysis by Auger and X-ray Photoelectron Spectroscopy, D. Briggs, J.T. Grant Eds., IM Publications, Chichester, 2003: pp. 377–389.

[24]   A. Arranz, C. Palacio, Factor analysis, a useful tool for solving analytical problems in AES and XPS: A study of the performances and limitations of the indicator function, Surf. Interface Anal., 22 (1997) 93–97.

[25]   R.E. Watson, Improved dynamic range and automated lineshape differentiation in AES/XPS composition versus depth profiles, Surf. Interface Anal., 15 (1990) 516–524.

5 Element Distribution With Depth

5.1. DEPTH PROFILE

Depth profiling is a special case of micro-local analysis providing the chemical composition along the depth direction. In depth profiling, x, y coordinates indicate the region where the signal is acquired while the z coordinate, perpendicular to the sample surface, indicates the sampling depth, namely the thickness of the overlayer where are generated the photoelectrons arriving to the analyzer. Generally the depth profile corresponds to the compositional analysis of thin sections of the specimen which are defined on a scale depth. Although X-photons can penetrate deeply into the materials inducing photoemission, only photoelectrons near to the surface may escape into the vacuum and be detected by the analyzer because of scattering processes. By definition, the sampling depth corresponds to three times the inelastic mean-free path (IMFP) λ (see Section 3.6.2). This value accounts for the 95.7% of photoelectrons generated near the surface and exiting into the vacuum with kinetic energy < 1 KeV. Then, 61.7% of them experience an interaction within λ. Depth information within a layer of thickness 3λ can be obtained in the angle-resolved XPS experiments where the sample is non-destructively probed. The value of λ depends on the nature of the sample and, remembering the expression of the *Universal Curve* (see equations 3.4 and 3.5), it is dependent on the material density, and on the photoelectron kinetic energy. Information on the sample composition in the bulk can be obtained only by eroding the sample surface thus allowing inner regions to be probed. In this kind of experiments, the surface is bombarded by ions, generally Ar^+ ions are used to sputter sample surface. Depending on the nature of the material, ion bombardment allows generation of profiles on depth of microns.

5.1.1. NON-DESTRUCTIVE DEPTH PROFILES

Non-destructive techniques used to obtain the material composition as a function of depth are all based on the property that the signal decays with depth. As an example, in Rutherford Back Scattering (RBS), a beam of He^+ primary ions are used to probe the material. The technique observes the secondary back scattered ions. Back scattering depends on the cross section of elements composing the material, while measuring the energy-loss, the depth information is obtained. Similarly, in Nuclear Reaction Analysis (NRA), primary ions, for example, N, accelerated at kinetic energies of the order of MeV, are able to react with the elements of the material depending on their reaction cross section. By increasing the energy of the primary beam, reactions occur at increasing depths. In this manner, NRA is capable of providing both the compositional and the depth information. Efficiency of both the techniques is linked to the

DOI: 10.1201/9781003296973-5

scattering cross section which is low for light elements. In this case, RBS is more appropriate for elements with atomic number higher than 10 although RBS cannot be utilized for heavy atoms because of the too high scattering cross section. NRA is used for light elements with $Z < 20$ and possessing a reasonable reaction cross section. With respect to these techniques, XPS provides information for all the elements of the periodic table. Also in this case, the efficiency of photoelectron generation is linked to the cross section of the elements, but this is not a substantial limit for the technique. Electrons generated by X-ray radiation can be used to map the depth distribution of elements by using:

- In angle-resolved XPS (AR-XPS), the sample is tilted with respect to the analyzer axis. In this case, by lowering the photoelectron take-off angle θ, the sampling depth scales with $3\lambda \sin \theta$ where λ represents the IMFP (see also Section 1.3.2). AR-XPS is generally carried out using hemispherical analyzers due to the possibility to select the collection solid angle by regulating the entrance iris to the analyzer and possibility to easily measure the take-off angle. The exact definition of both these parameters becomes very complex in the case of the cylindrical mirror analyzer. This is due to the geometrical reason, since in this case, the collection angles are defined by the space between inner and outer cylinders of the analyzer as depicted in Figure 5.1a. This circular aperture cannot be modified, and since the radius of the cylinders is big (some centimeters), the evaluation of the inclination with respect to the analyzer axis becomes rather difficult as anticipated. Electrons emerging from the surface with the same take-off angle with respect to the surface normal n arrive to the CMA with rather different directions with respect to the analyzer axis. This is shown in Figure 5.1a where electrons exiting along the surface normal n are differently tilted with respect to the acceptance cone edges A, B lying in a vertical plane. The problem can be solved using a modified double pass analyzer equipped with a rotatable drum with an aperture which is used to select a well-defined emission direction θ_e, φ_e from the total acceptance cone of the CMA. This experimental configuration render possible to maintain fixed the orientations of the X-source and of the sample with respect to the analyzer axis and select the emission angle modifying the position of the rotatable drum aperture [1].

The symmetry of the electron trajectories inside the CMA makes electrons entering and exiting the analyzer to have the same conical distribution. Referring to Figure 5.1b, the geometry is described using the angles θ_e and φ_e for the polar and azimuthal angles of the emitted photoelectron with respect to the x, y coordinate system aligned with the sample surface. R and φ polar coordinates are used in the plane of the detector. The angles are related via the equations

$$\cos \theta_e = -\sin \alpha \, \sin\left(42.5^0\right) \, \sin \varphi + \cos \alpha \, \cos\left(42.3^0\right)$$

$$\sin \varphi_e \sin \theta_e = \sin\left(42.3^0\right) \cos \varphi$$

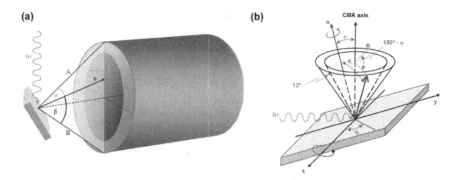

FIGURE 5.1   (a) Electrons emitted along the sample normal *n* are tilted by α with respect to the upper edge A of the analyzer cone and β with respect to B making definition of the sample inclination not possible with a common CMA. (b) Geometry of the sample surface coordinate system and the CMA acceptance cone. The sample is rotated by α around the x axis. θ_e and φ_e are exit angles at the sample. R and φ are the coordinates in the detector plane.

When α = 0^0, θ corresponds to the CMA aperture (typical angle = 42.3°) for every φ and only azimuthal dependence is left. For α = 42.3°, the surface normal is aligned along the analyzer axis, and almost the full polar range is accessible using the rotation around the surface normal. For more details, see [2].

- Energy-Resolved XPS (ER-XPS) is realized by varying the X-photon energy and maintaining the position of the sample fixed with respect to the analyzer axis. This method requires the XPS instrument be equipped with different X-source anodes. As an example, Al and Ag anodes produce Al Kα and Ag Lα radiations of energies respectively equal to 1486.6 eV and 2984.3 eV. ER-XPS is also performed using synchrotron radiation where it is possible to select the desired photon energy in a continuum range from few eV to tens of KeV. Because the IMFP depends on the electron kinetic energy, using different X-photon energies is possible to vary the thickness of the sampled region, thus obtaining the element distribution with depth.

In Figure 5.2a, it is shown the typical change of the spectra from a naturally oxidized silicon sample, by varying the take-off angle. Spectra are normalized to the bulk Si° component at 99.5 eV. It can be observed that decreasing the take-off angle (i.e., increasing the inclination with respect to the vertical axis), there is an increase in the oxidized silicon components in the range energy 101–105 eV. This mirrors the higher surface sensitivity resulting in a higher abundance of oxygen. This picture is confirmed by Figure 5.2b displaying the concentration profile of the chemical elements detected in the analyzed sample. Carbon is present as a surface contamination. The depth profile shows that at take-off = 90° corresponding to the higher sampling depth, the concentration of Si is higher. By decreasing the value of the take-off angle till 15°, the concentration of Si decreases and is compensated by

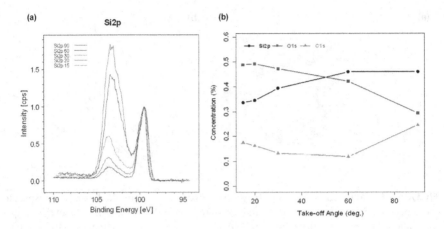

FIGURE 5.2 (a) Trend of the Si 2p Core-Line as a function of the take-off angle. (b) Correspondent change on the abundance of Si, O, and C elements detected on the sample surface.

a correspondent increase of the oxygen abundance. At take-off = 15°, the trend of the oxygen concentration flattens due to the increase of the carbon concentration. However, as shown in Figure 5.2a, the intensity of the oxidized Si 2p components is higher at the lower angles.

Another example of AR-XPS is reported in [3], where authors studied the structure of TiN/HfO$_2$ nanofilms grown on In$_x$Ga$_{1-x}$As substrates. First, the InGaAs substrate was cleaned with an H$_2$O:HF (100:1) solution to remove any contamination. Then HfO$_2$ was grown in an atomic layer deposition chamber. Finally, for the fabrication of the MOS structure, a thin TiN layer was deposited by sputtering. The TiN layer is important to avoid desorption of oxygen from HfO$_2$ during the annealing performed at 500 °C and 700 °C. The MOS structure was analyzed through AR-XPS which made possible to characterize the various layers treated at different temperatures. The authors applied the MultiLayer Model (MLM) to the spectral data to extract the structure of the films namely the thickness and composition of the component layers [4]. This DFT method, similar to that described in [5], allows estimating the transport activation energy of In through hafnia, thus solving the controversial results about out-diffusion of In [6] or As into HfO$_2$ [7]. Figure 5.3 put in evidence the variation of the Ti 3p, As 3d, In 3d, Ga 3p, O 1s, and N 1s core-lines with the take-off angle. To evaluate the trend with variation of the take-off angles, spectra were fitted using Voigt functions. Authors found that experimental and theoretical data indicate the presence of As^0 at the interface. The trend of N with the take-off angle and its BE are in agreement with the component of Ti at 34.6 eV and is explained with a layer of TiN located at the top of the hafnia. As for Ga, it is in the form of gallium oxide and spectra show that it is distributed at the oxide/semiconductor interface. Differently, spectra show that with temperature, it migrates and diffuses into the Ti layer. Finally, presence of C is compatible with adventitious carbon at the sample surface.

As observed, the element distribution with depth can also be reconstructed by varying the photon energy and consequently the sampling depth. This is shown in

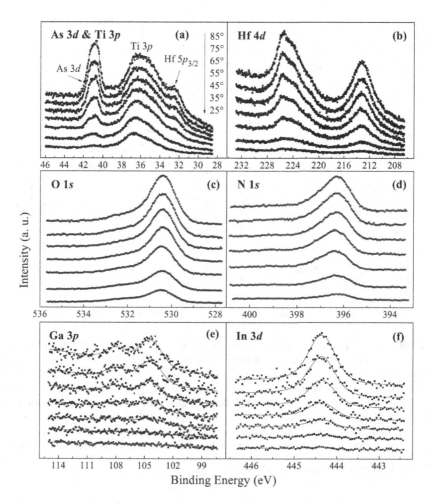

FIGURE 5.3   Angle-resolved XPS spectra (dots) and fits (lines) for the as-deposited sample.

Source: Reprinted with permission from [3]

Figure 5.4 where synchrotron radiation is used to perform ER-XPS in ultra-thin silane films [8]. Authors deposited two different ultrathin organic monolayers made of amino- or benzamido-silane molecules on a silicon oxide surfaces. Thanks to the exceptional surface sensitivity of the apparatus, authors were able to obtain a reliable chemical-in-depth information with a z-resolution of about 1 nm. Figure 5.4a, b shows the change of the intensity of the Si 2p spectral features respectively of amino- or benzamido-silane molecule layers by varying the excitation energy hv from 21 eV to 1,487 eV. It may be appreciated that at the lower excitation energies hv, the Si 2p spectra show prominent components at ~104 and 102 eV deriving from the deposited organic layers. By increasing the photon energies, also the elemental Si^0 deriving from bulk Si atoms appears revealing an increasing sampling depth. Possessing information

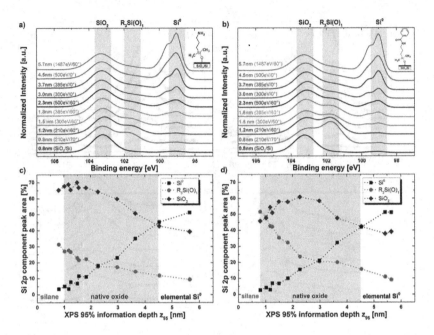

FIGURE 5.4 Top row: Si 2p synchrotron XPS data of aminosilane (a) and amidosilane (b) monolayers on SiO₂/Si substrates at different excitation energies (hv) and electron-emission angles (θ) presented as a non-destructive chemical depth profile versus the XPS 95% information depth z95 ranging from 0.8 nm to 4.5 nm. Laboratory XPS data with z95 = 5.7 nm at hv = 1486.7 eV and θ = 60° were included to expand the z range. Also an oxidized Si wafer (SiO/Si) is shown as blank. Bottom row: in-depth distribution of Si2p components Si⁰ (elemental), R₃SiO₁ (silane), and SiO₂ (native oxide) obtained from the Si 2p core-level spectra at different z values of aminosilane (c) and amidosilane (d) monolayers after curve fitting.

Source: Reprinted with permission from [8]

about the material density, authors were able to estimate the sampling depth as a function of the photon energy. Spectra show that the Si 2p component at 99 eV derives from photoelectrons generated at a depth ≥ 4.5 eV. For z values ranging from 4 to 1.5 nm, organic (silane)–inorganic (SiO₂/Si) interface is probed. In particular, increasing the sampling depth, the Si 2p core-level spectra are more and more dominated by the SiO₂ component at a binding energy of 103 eV deriving from the native oxide layer [9]. Decreasing the photon energy below 500 eV at third Si component placed at a binding energy of 101.7 eV starts to appear. This component derives from the organic Si atoms of the R₃SiO₁ silane overlayer. The observed component is assigned to the Si-O-Si single bond in siloxane [10–12]. The intensity of this component increases by decreasing hv testifying that the organic layer is characterized by a thickness of about 1 nm. In the case of benzamido-silane (Figure 5.4b), the siloxane component at 101.7 eV becomes the dominant species in the XPS spectra. The higher peak area of this component compared to the correspondent in the amino-silane sample was corroborated by a higher nitrogen amount for amidosilane monolayers as estimated by XPS. The reason might be (i) the benzamide group prevents side reactions of free amines; (ii) and/or presence of intermolecular hydrogen-bonding between adjacent amide leads to a higher silane surface density [13]. In Figure 5.4b, the second row displays the concentration profile of the Si components associated to SiO_2, organic Si deriving from the silane molecules and bulk Si^0. As expected, organic Si from the deposited amino- or amido-silane on the surface and Si from bulk have opposite trends, the first decrease with increasing sampling depth. The oxidized Si interface placed between these two materials is also mirrored by the trend of the depth profiles, since it has a maximum located below the surface top layers as expected.

5.1.2. DESTRUCTIVE DEPTH PROFILE

As seen in the previous section, profiling the top layers of sample is possible by simply modifying the inclination of the sample with respect to the analyzer axis or changing the excitation photon energy. However, the drawback is the limited sampling depth which limits the analysis to the surface. If information about the bulk composition is required, the only solution is to erode the surface. This operation is made by bombarding the surface with energetic primary particles, usually Ar ions, with kinetic energy between 0.5 keV and 5 keV range. Increasing the sputtering time increases the erosion, and layers beneath the original surface are gradually exposed for the analysis.

Destructive erosion of the surface is used in two different ways to reconstruct the in-depth distribution of composition. Instruments as the Secondary Ions Mass Spectrometers (SIMS) or Secondary Neutral Mass Spectrometers (SNMS) analyze, respectively, the ionized or neutral fragments generated by the bombardments. Differently, XPS, Auger Electron Spectroscopy (AES), and Ion Scattering Spectroscopy (ISS) analyze the surface left after the sputtering. These two approaches markedly differ in terms of elemental composition gained, sensitivity, dynamic range, and information depth. The sputtering process is itself, however, common to two analysis methods and is therefore considered as a separate physical process. Ideally, ion bombardment in controlled conditions can remove atoms mainly from the first monolayer of a solid [14, 15]. Then, a depth resolution in the monolayer range should be obtained. Unfortunately sputtering does not proceed through a layer-by-layer removal: erosion is the result of a complex interaction of the ion beam with the sample surface. A variety of perturbations like deformation of the original morphology, changes of the sample composition introduced by the sputtering, Ar-ion implantation cause a profile broadening and shape changes [16]. This has led to specific studies for a better understanding of the main physical processes involved in sputter erosion to enable optimized profiling conditions for the achievement of high depth resolution [17].

In XPS and AES instruments used for depth profiling, an important element is the source of ions with energy in the range 0.5–5 KeV, utilized to sputter the sample surface. Modern instrument are equipped with differentially pumped ion-gun sources, allowing the pressure in the analysis chamber to be in the UHV regime. Ion-guns are also capable to produce focused ion beams rasterized on the sample surface to produce a crater. The ion gun is fixed to the analysis chamber in a well-specific geometry defining the incidence angle with respect to the normal to the specimen surface. The beam energy can be selected and chosen with regard to the hardness of the specimen.

Increasing the beam energy results in a high etching rate and is applied to rough samples while low beam energy is generally used in the case of flat surfaces.

To ensure a constant sputtering rate, essential is that the ion beam parameters are constant during the sputtering time. An argon leak valve is used to ensure a constant flux and a constant ion beam current. Depending on the ion flux, their energy, and the nature of the specimen, the sputter rate can range from less than a monolayer per minute to values higher than 10 nm/min. The erosion process is well understood in the case of monoatomic ion impacts. Sputtering derives from a sequence of elastic collisions between impinging ions and sample atoms. This process is generally referred to as *collision cascades*, where elastic scattering events generate the erosion. To improve the uniformity of the sputtering, many instruments offer the possibility to rotate the sample around the vertical axis (azimuthal sample rotation) [18, 19]. The sputter depth profiling consists in acquiring the photoelectron spectra of the required elements as a function of the sputtering time. Given a constant ion energy and beam current density, it may be assumed that the sputtered depth is proportional to the sputtering time for a homogeneous sample composition. Then, a sputter depth profile experiment needs: (i) optimization of the correct ion beam current and energy; (ii) selection of the irradiation position, which is essential for non-homogeneous specimens; (iii) ion beam scanning X, Y width determining the amplitude of the crater which should be sufficiently extended to ensure a flat bottom [20]. Table 5.1 lists the sputtering yield of some elements and oxides [21–24]:

TABLE 5.1

Sputtering Yield (atoms/ion) of Selected Elements at Increasing Atomic Number Z for Ar⁺ Ions. Values Are Referred to 1 KeV Ion Energy and 45° Incidence Angle

Element or oxide	Sputtering yield [1 keV] (atoms/ion)	$M/(n\rho)$ (cm³/g-atom)	Sputtering rate ratio to Ta_2O_5 (1 keV) (*)
^{6}C	0.98	5.21	0.31
^{12}Mg	4.90	14.0	4.5
^{13}Al	2.71	10.0	1.8
^{14}Si	1.63	12.0	1.3
^{22}Ti	1.55	10.55	1.1
^{24}Cr	2.77	7.23	1.3
^{26}Fe	2.81	7.08	1.3
^{27}Co	3.00	6.62	1.3
^{28}Ni	3.08	6.60	1.3
^{29}Cu	4.00	7.09	1.9
^{31}Ga	3.57	11.43	2.7
^{32}Ge	2.32	13.65	2.1
^{33}As	10.0	13.07	8.6
^{41}Nb	1.65	10.48	1.2
^{42}Mo	2.15	9.39	1.3
^{46}Pd	4.62	8.87	2.7

Element or oxide	Sputtering yield [1 keV] (atoms/ion)	M/(nρ)(cm³/g-atom)	Sputtering rate ratio to Ta₂O₅ (1 keV) (*)
^{47}Ag	5.77	10.27	3.9
^{49}In	4.56	15.70	4.7
^{72}Hf	1.75	13.40	1.5
^{73}Ta	1.60	10.87	1.1
^{74}W	1.62	9.53	1.0
^{78}Pt	2.80	9.10	1.7
^{79}Au	4.19	10.21	2.8
^{82}Pb	4.94	18.23	5.9
^{83}Bi	5.39	21.43	7.5
Al_2O_3	1.8	5.13	0.62
SiO_2	2.7	7.58	1.35
TiO_2	1.8	6.30	0.73
Cr_2O_3	2.0	5.83	0.72
Fe_2O_3	2.0	6.10	0.82
Fe_3O_4	2.5	6.39	1.03
ZnO	3.0	7.25	1.41
CeO_2	1.4	7.50	0.70
HfO_2	1.5	7.25	0.72
ITO	2.54	7.68	1.41
Ta_2O_5	1.97	7.70	1.00

An example of XPS sputter depth profiling is given in [25] where authors studied the effect of temperature on a Mo-C-Mo and Si-C-Mo and Mo-C-Si multilayers. Mo/Si multilayer structures as are widely studied as Bragg reflectors for Extreme Ultra-Violet radiation. However to preserve the reflection properties of the multilayered structure, important is to avoid any kind of element interdiffusion caused by the thermal load during deposition. To reduce interdiffusion, generally a barrier layer is used. The barrier should possess high chemical stability and low miscibility in Mo and Si matrixes and should maintain or possibly improve the contrast among Mo and Si layers and the whole reflectivity. A nanometer-thick carbon layer was utilized as a barrier to prevent interdiffusion. To study the effect of the C layer, authors first studied the Mo-C-Mo and Si-C-Si multilayers and the barrier effect of the C layer with the temperature. Figure 5.5 shows the results.

One hour of annealing at 500 °C causes interdiffussion of Mo and C. Differently, even at 600 °C, no interdiffusion of Si and C was observed (Figure 5.5b). Therefore, the presence of a C layer between Mo and Si is expected to generate a strong diffusion asymmetry. This effect is shown in Figure 5.6, where annealing was applied to Si(wafer)-Si-C-Mo and to a Si(wafer)-Mo-C-Si multilayers. For the as-deposited structures, Figure 5.6a shows a broader interface for Mo deposited on C on Si with

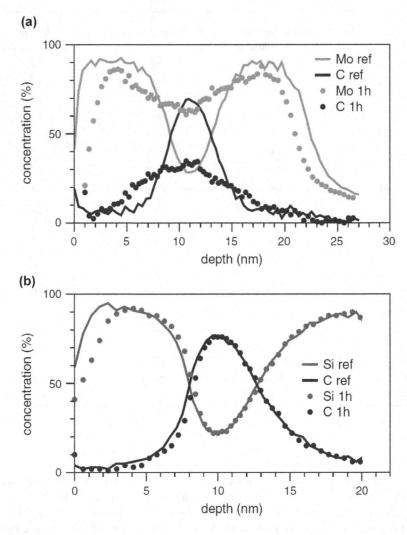

FIGURE 5.5   XPS sputter depth profile of: (a) Mo/C/Mo before (solid line) and after (dotted line) annealing at 500 °C for 1 h; (b) Si/C/Si before (solid line) and after (dots) annealing at 600 °C for 1 h.

Source: Reproduced with permission from [25]

regard to Si on C on Mo. This is interpreted as a highly localized depth distribution of C with formation of a thinner carbide interface.

The annealing at 600 °C for 3 h induces a complete interdiffusion of the C into the Mo layer in the Si (wafer)/Si/C/Mo structure (Figure 5.6a). Same fate occurs in the Si(wafer)/Mo/C/Si multilayer. Also in both the cases, the Mo interface, which is initially sharp, broadens extending into the Si layer. However, authors observed that Mo_2C/Si multilayer structures display a superior thermal stability up to 500 °C with

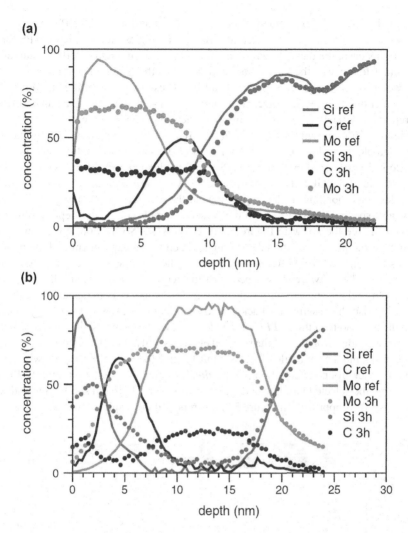

FIGURE 5.6 XPS sputter depth profile of: (a) Si(wafer)/Si/C/Mo before (solid line) and after (dots) annealing at 600 °C for 3 h; (b) Si(wafer)/Mo/C/Si before (solid line) and after (dots) annealing at 600 °C for 3 h.

Source: Reproduced with permission from [25]

respect to the Mo/Si system. A perfect stability of the multilayer is obtained by alternating Si and crystalline molybdenum carbide Mo_2C. The formation of the carbide renders carbon and Mo stable also after 5 h at 600 °C annealing.

5.1.3. DEPTH PROFILE OPTION

The *Depth Profile* option is part of the *Analysis* menu. This option is used to analyze one XPS Sample data file or multiple acquisitions obtained from an ARXPS or a sputter depth-profile experiment. Let us start considering an ARXPS experiment. The user can store spectra acquired at different tilt angles, in one unique XPS Sample. In this case, the data file will contain a wide spectrum and a certain number of core-lines for each of the sample inclinations. The other solution is to record spectra for each given tilt angle in separate data files. These two possibilities hold also for the sputter depth profiles, where, at the end of each sputtering cycle, acquisitions are stored in individual files or saved always in the same XPS Sample.

Figure 5.7a represents the *Depth Profile* GUI. In this example, a list of separated XPS Sample data files are recorded for each of the sample inclinations and must be loaded in *RxpsG*. As shown in Figure 5.7a, all the loaded XPS Samples are listed on the left side of the GUI. The depth-profile analysis starts with the choice of the XPS Samples corresponding to the acquisitions at the different take-off angles. The user must also specify if ARXPS (as in the present example) or sputter depth profile was performed (see the arrow in Figure 5.7a). Upon ARXPS selection, a window opens asking for the number of the take-off angles used in the experiment and the relative values (see Figure 5.7b). After angle definition, the window is closed by pressing the *OK* button. The software shows the list of the acquired core-lines and the user must select which elements want to profile, as Si 2p, O 1s, and C 1s in this example in Figure 5.7c. Also the baseline for background subtraction (Shirley in the example) must be indicated and then the *SELECT/ADD BASELINE* button pressed to proceed. In the graphical window, for each of the selected elements will appear the core-lines acquired at the different take-off angles. The user must define the ends of the region where perform the background subtraction (the *RegionToFit* for instance) and promptly *RxpsG* will apply the selected baseline and will perform the background subtraction and will ask for analysis approval. In Figure 5.8, it is reported the result for the Si 2p core-line.

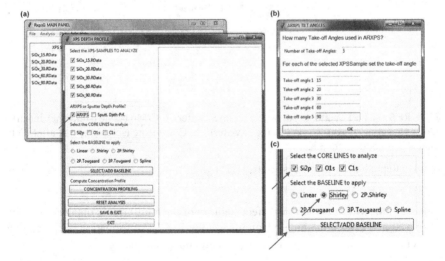

FIGURE 5.7   (a) The *Depth Profile* GUI. On the top-right part are indicated all the XPS Samples loaded into *RxpsG*. In this example, individual XPS Sample files are saved for each of the tilt angles. (b) For each of the chosen XPS Samples, the user must specify the *take-off* angle. (c) To accomplish the analysis, a baseline must be selected for background subtraction.

When the baseline is applied to all the element's core-lines, simply pressing the button *CONCENTRATION PROFILING*, the element distribution as a function of the take-off angle is obtained. The result is displayed in Figure 5.9. As it appears, the correspondent element concentrations are reported in the textual window of the right side of *Depth Profile* GUI. The data can be selected and copied in the desired document.

As a second example, let us analyze a sputter-depth profile saved in one unique file. Figure 5.10 shows the GUI where the single file WCrY_470DP.Rdata, the *Sputter Depth Profile* mode, and the C 1s, O 1s, Cr 2p, and Y 3d elements are selected for the profile. Also in this case, a *Shirley* baseline was chosen for background subtraction.

As in the previous case for each of the selected elements, the *RxpsG* will ask to identify the edges of the region where to apply the baseline. When background subtraction is performed for all the acquired core-lines, the depth concentrations as

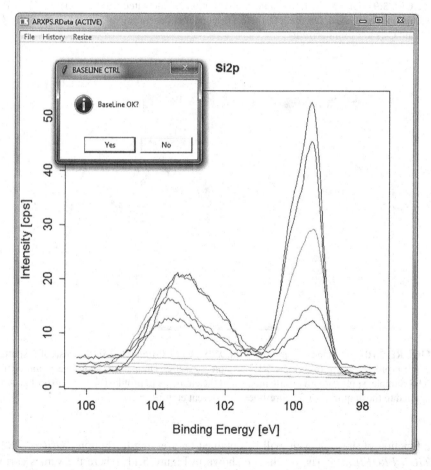

FIGURE 5.8 Following selection, a Shirley baseline was applied to all the Si 2p core-line acquired at the different take-off angles.

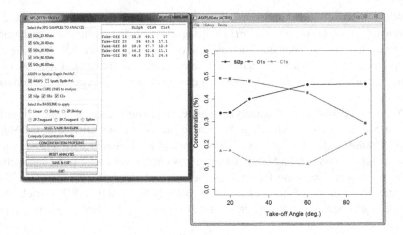

FIGURE 5.9   On the right, the element distribution as calculated from the series of ARXPS acquisitions is plotted. On the left, the correspondent concentrations relative to the Si 2p, O 1s, and C 1s are summarized.

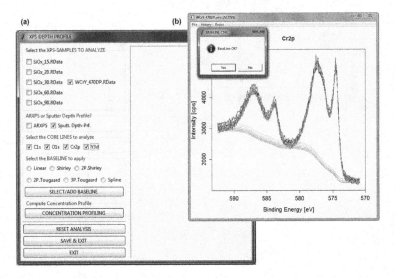

FIGURE 5.10   (a) Selected options in the *XPS Depth Profile* GUI in the case of a sputter depth profile saved in one single XPS Sample data file. The picture shows selection of the XPS Sample, sputter depth profile mode, and the elements to profile. (b) The *Shirley* baseline applied to the acquired Cr 2p core-lines at different etching cycles.

a function of the etch cycle will be obtained by pressing the button *CONCENTRA-TION PROFILING*. The results are shown in Figure 5.11, where the values corresponding to the plotted data are reported in the text window on the right side of the *Depth Profile* GUI.

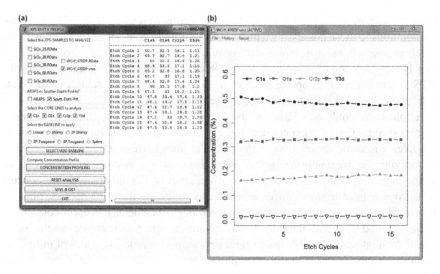

FIGURE 5.11 (a) are plotted the element distribution as calculated from the sputter depth profile acquisition. (b), the correspondent concentrations relative to the C 1s, O 1s, Cr 2p, and Y 3d are summarized.

5.2. MAXIMUM ENTROPY METHOD

XPS is a technique widely utilized for analyzing the outermost atomic layers of solid specimens and for gaining their composition with depth by variation of the sample's tilt angle. Obtaining this information is possible, because the attenuation of the outgoing photoelectron intensity of a given element vary with the depth and with its distribution with depth. Because a variation of the take-off angle results in a change in the sampling depth, spectra acquired at different emission angles contain information on the sample structure. So, the use of angle-dependent XPS measurements for the non-destructive determination of the element concentration profiling is considerable. Numerous algorithms been reported to retrieve this information [26–31]. Common problem to these methods is the extremely high sensitivity of the estimated depth concentration profiles to errors superposed to experimental data [29, 32]. This originates in the case of inverse problems where initial conditions are deduced from final data and generate an ill-posed problem. In our case the measured angle dependent intensities are used to estimate the depth distribution of elements. The element distribution with depth cannot be simply obtained from simple inspection of ARXPS spectra. Acquisitions at five or six angles are commonly done to reconstruct depth compositional variations. Noise is overlapped to the spectra and moreover, only a weak contribution is expected from deep layers. Other problems derive from the sample nature affecting the elastic scattering and electron refraction and from the surface roughness. Then small inaccuracies in the data can result in consistent changes in the estimations and a simple minimization of a sum of squares error between the calculated and measured data is not sufficient to correctly define the sample structure. A possible solution is to introduce constraints deriving from prior knowledge of the sample composition and some form of regularization to restrict the number of possible solutions to those possessing a higher verisimilitude. In practice, these constraints result in restrictions on samples with only a small number of components or with slowly varying compositions with depth. Other approaches attempted to estimate the XPS spectral intensities on the basis of model structures of the samples and iterate until the calculated intensities matched the original data [29, 33–35]. However, in principle, there could be a large number of possible models reproducing the experimental spectra. Only the experience of the operator may decide which of the models provides a realistic description of the data. An alternative solution to this problem is applying the *Maximum Entropy Method*. Entropy is a physical entity associated to the degree of disorder, randomness, or uncertainty. For this reason, the concept of entropy is used in many different areas encompassing thermodynamics, physics, statistical physics, chemistry, biology, in cosmology, weather science, economics and sociology, and information theory. L. Boltzmann interpreted the entropy as the measure of the number of possible microstates (in thermodynamics the states of individual atoms/molecules) of a system determining the macroscopic condition of that system:

$$S_B = k_B \log W \qquad 5.1$$

where S_B denoted the entropy, k_B is the Boltzmann constant, and W represents the system multiplicity, that is, the number of microstates corresponding to a macroscopic thermodynamic system.

A statistical version of equation (5.1) was introduced by Shannon in 1948 [36]

$$S = \sum_i p_i \log(p_i) \qquad 5.2$$

where p_i is the probability of the system to be in the state *i*. In information theory, equation (5.2) tells us that the entropy S associated to a given variable describes the level of "information" or "uncertainty" characterizing the possible outcomes of that variable. In data analysis, the entropy assumes the meaning of absence of information content in a reconstruction. In ARXPS, we would like to reconstruct the sample structure from angle dependent noisy spectra. The idea underlying the maximum entropy method is the minimization of the information needed for the reconstruction since it is not needed to model the spectral noise. The solution of the problem is then the one possessing the minimum information (i.e., maximum entropy) reproducing the spectral data. In practice, the reconstruction obtained maximizing the entropy is the simplest solution among the possible alternatives which would produce statistically equivalent data considered variations due to the spectral noise. Application of the maximum entropy to element distribution is due to Skilling and Gull [37]

$$S = \sum_j \sum_i n_{j,i} - m_{j,i} - n_{j,i} \log(n_{j,i} / m_{j,i}) \qquad 5.3$$

Here $n_{j,i}$ is the concentration of element *j* at depth *i*; $m_{j,i}$ is the initial estimate of the abundance of element *j* at depth *i*. Concentrations are forced to be positive because of the logarithm. The solution to find the depth distribution of elements in a specimen was initially computed by maximizing S in equation (5.3) with respect all the possible element distribution $n_{j,i}$. This condition is obtained by verifying that the weighted square error χ^2 is consistent with the data uncertainty

$$\chi^2 = \sum_k \left(I_k^{Calc} - I_k^{Obs} \right)^2 / \sigma_k^2 \leq N \qquad 5.4$$

I_k^{Calc} and I_k^{Obs} are the calculated and observed photoelectron intensities at the k_{th} angle; σ_k^2 is the variance of the k_{th} measurement, while N is the number of independent observations in the data. Equation (5.4) provides an expression for the entropy which acts as a regularization function. Because the entropy is non-linear, the maximization process is iterative. At beginning, the value of the entropy S is computed for the $n_{j,i}$ corresponding to the $m_{j,i}$ experimental values. Then $n_{j,i}$ are modified to reduce the weighted sum of square errors χ^2 until it corresponds to the number of independent observations N. For determining the maximum of the entropy the reduction of the χ^2 may be coupled to the maximization of the function

$$Q = \alpha S - \chi^2 / 2 \qquad 5.5$$

where α is a regularizing parameter which is reduced until the condition $\chi^2 = N$ is reached.

5.2.1. IMPLEMENTATION OF THE MAXUMUM ENTROPY METHOD TO THE ANALYSIS OF ARXPS SPECTRA

The description of the *Maximum Entropy* method used to determine the depth distribution of element in a material is based on the work of Livesey and Smith [38]. As described in Section 4.1, the general expression used to compute the element concentration is (equation 4.7)

$$X_a = I_a / R_a \bigg/ \sum_i I_i / R_i \qquad 5.6$$

Here X represents the atomic composition of element *a* in a sample containing *i* chemical species, I_a is the spectral intensity corresponding to *a*, while R_a represents the sensitivity factors of *a* or *i* elements. *R* is defined in Section 4.1. Generally *R* accounts also for instrumental factors (the efficiency of the analyzer and of the detector) and are provided by the instrument manufacturer to obtain reasonable values of the element abundances in homogeneous samples. On the contrary, if the composition changes with depth, the estimation of the element concentrations may suffer from significant errors. In Section 1.3.2, it was described the variation of the signal intensity as a function of the depth z following a Beer-Lambert law:

$$I = I(z) \exp(-z / \lambda \cos \theta) \qquad 5.7$$

where λ represents the attenuation length of the specimen and θ is the photoelectron emission angle measured with respect to the sample normal. The objective is to use this angle dependence of the element concentrations X_a to reconstruct the composition of the sample as a function of the depth z. Concentrations are used instead of intensities to separate the quantification process from that of structure determination. This approach is essentially the same described in [39] but now based on atomic concentrations. The use of angle dependent concentrations allows for analyzing the effect of uncertainties in the quantification process on the determination of sample structure. In addition, the use of concentrations directly adds the constraint that the abundances of all the elements at each depth should sum to 1.0 (i.e., sum of concentrations must give the 100%). To estimate how the element depth distribution in a specimen influences the XPS spectral intensities we refer to the scheme shown in Figure 5.12. The sample is divided in layers each containing a certain number of elements a, b, c, . . ., etc. with concentrations x_a, x_b, x_c. . ., etc. $n_{j,i}$ is the proportion of element j in the i-th layer; it is the fraction composition of element j in the layer i so that summing over all layers and all element, we must obtain

$$\sum_{j=1}^{M}\sum_{i=0}^{N} n_{j,i} = 1 \qquad\qquad 5.8$$

To each layer j is associated a transmission function $T_j(\theta)$ defined by

$$T_j(\theta) = \exp(-t / \lambda_j \cos\theta) \qquad\qquad 5.9$$

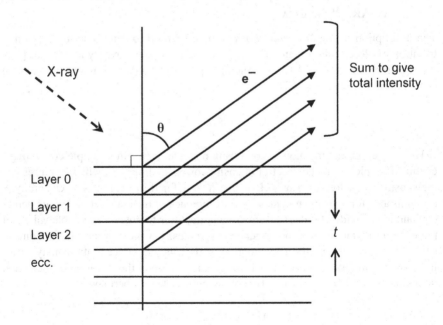

FIGURE 5.12   The structure used to model XPS intensity variations in layered samples.

Source: Reprinted with permission from [38]

Here λ_j represents the attenuation length of the photoelectrons generated by the element j.

For any element j, the photoelectron intensity from the sample is obtained by summing the contribution over all layers. If we normalize to unity the intensity of the X-ray excitation

$$I_j(\theta) = k_j[n_{j,0} + n_{j,1}T_j(\theta) + n_{j,2}T_j(\theta)^2 + n_{j,3}T_j(\theta)^3 + \ldots]$$ 5.10

or introducing the summation

$$I_j(\theta) = k_j \sum_{i=0}^{N} n_{j,i} T_j(\theta)^i$$ 5.11

The cross sections of the different elements j are described by the constant k_j. The relative sensitivity factor R_j for the element j is then described by

$$R_j = k_j \sum_{i=0}^{N} T_j(\theta)^i$$ 5.12

The total signal intensity from all elements generated by the X-excitation is obtained by summing all the intensities from each of the elements j

$$I_T(\theta) = \sum_{j=1}^{M} k_j \sum_{i=0}^{N} n_{j,i} T_j(\theta)^i$$ (5.13)

Now, for each elements $a, b, c, \ldots$, the relative concentration expressed as a function of emission angle will be obtained by substituting equations (5.11) and (5.12) in equation (5.6). Note that the constants of proportionality k_j for each element cancel and therefore the concentration of the element j is expressed by

$$X_j(\theta) = \frac{\sum_{i=0}^{N} n_{j,i} T_j(\theta)^i \bigg/ \sum_{i=0}^{N} T_j(\theta)^i}{\sum_{j=1}^{M} \left(\sum_{i=0}^{N} n_{j,i} T_j(\theta)^i \bigg/ \sum_{i=0}^{N} T_j(\theta)^i \right)}$$ 5.14

Equation (5.14) defines the forward transform which describes the theoretical XPS spectral intensities of elements deriving from any arbitrary layered model structure as a function of the emission angle θ.

The reconstruction of the depth profile of elements is then computed by estimating the concentrations of all elements j at angle θ, $X_j(\theta)$, by applying equation (5.14), where $n_{j,i}$ are provided as estimates. These values are then used to generate, for each element and angle θ, a spectral intensity which is compared with the experimental data. An error is then computed applying equation (5.4). The estimates of the element concentrations $n_{j,i}$ are then modified in order to maximize the entropy S and minimize the error χ^2, that is, maximize the parameter Q defined in equation (5.5).

The computation of the element profile using the *Maximum Entropy* method was implemented but not included in the *RxpsG* software since still requires testing and a careful control of the results provided. There is a wide literature on the application of the maximum entropy method. The reader can also refer to published papers [40–44]. References [45] and [46] provide a statistical support to the *Maximum Entropy Method* if the reader is interested to deepen the theoretical basis of this technique. The maximum Entropy method was developed by the authors by still not inserted in the *RxpsG* software because the procedure is under careful testing.

REFERENCES

[1]   S. Hofmann, J.M. Sanz, Characterization of Contamination Layers by Emission Angle Dependent XPS with a Double-pass CMA, Surf. Interface Anal., 6 (1984) 75.

[2]   A. Bosch, H. Feil, G.A. Sawatzky, A simultaneous angle-resolved photoelectron spectrometer, J. Phys. E, 17 (1984) 1187–1192.

[3]   A. Sanchez-Martinez, O. Ceballos-Sanchez, M.O. Vazquez-Lepe, T. Duong, R. Arroyave, E. Espinosa-Magana, A. Herrera-Gomez, Diffusion of In and Ga in TiN/HfO2/InGaAs nanofilms, J. Appl. Phys., 114 (2013) 143504-1–143504-6.

[4]   A. Herrera-Gomez, Self Consistent ARXPS analysis for multilayer conformal films with abrupt interfaces, Internal Report, Queretaro, Mexico, http://www.qro.cinvestav.mx/~aherrera/reportesInternos/arxpsAnalysisSharpInfefaces.pdf, 2008.

[5]   O. Ceballos-Sanchez, A. Sanchez-Martinez, M.O. Vazquez-Lepe, T. Duong, R. Arroyave, F. Espinosa-Magana, A. Herrera-Gomez, Mass transport and thermal stability of TiN/Al2O3/InGaAs nanofilms, J. Appl. Phys., 112 (2012) 053527.

[6]   H.-D. Trinh, Y.-C. Lin, H.-C. Wang, C.-H. Chang, K. Kakushima, H. Iwai, T. Kawanago, Y.-G. Lin, C.-M. Chen, Y.-Y. Wong, G.-N. Huang, M. Hudait, E.Y. Chang, Effect of postdeposition annealing temperatures on electrical characteristics of molecular-beam-deposited HfO2 on n-InAs/InGaAs metal-oxide-semiconductor capacitors, Appl. Phys. Express, 5 (2012) 021104.

[7]   R. Suri, B. Lee, D.J. Lichtenwalner, N. Biswas, V. Misra, Electrical characteristics of metal-oxide-semiconductor capacitors on p-GaAs using atomic layer deposition of ultrathin HfAlO gate dielectric, Appl. Phys. Lett., 93 (2008) 193504.

[8]   P.M. Dietrich, S. Glamsch, C. Ehlert, A. Lippitz, Synchrotron-radiation XPS analysis of ultra-thin silane films: Specifying the organic silicon, Appl. Surf. Sci., 363 (2016) 406–411.

[9]   M.P. Seah, S.J. Spencer, Ultrathin SiO2 on Si II issues in quantification of the oxide thickness, Surf. Interface Anal., 33 (2002) 640–652.

[10]   R.A. Shircliff, P. Stradins, H. Moutinho, J. Fennell, M.L. Ghirardi, S.W. Cowley, H.M. Branz, I.T. Martin, Angle-resolved XPS analysis and characterization of monolayer and multilayer, Langmuir, 29 (2013) 4057–4067.

[11]   M.R. Alexander, R.D. Short, F.R. Jones, W. Michaeli, C.J. Blomfield, A study of HMDSO/O_2 plasma deposits using a high-sensitivity and -energy resolution XPS instrument: Curve fitting of the Si 2p core level, Appl. Surf. Sci., 137 (1999) 179–183.

[12]   T. Gross, D. Treu, W. Unger, Standard operating procedure (SOP) for the quantitative determination of organic silicon compounds at the surface of elastomeric sealants, Appl. Surf. Sci., 179 (2001) 109–112.

[13]   M.A. Ramin, G. Le Bourdon, K. Heuze, M. Degueil, T. Buffeteau, B. Bennatau, L. Vellutini, Epoxy-terminated self-assembled monolayers containing internal urea or amide groups, Langmuir, 31 (2015) 2783–2789.

[14] P. Sigmund, Sputtering by Ion Bombardment: Theoretical Concepts, in: Sputtering by Particle Bombardment I—Topics in Applied Physics, 1st Edition, Springer, Berlin; Heidelberg, 1981.

[15] K. Wittmaack, Surface and Depth Analysis Based on Sputtering, in: Sputtering by Particle Bombardment III, R. Behrisch, K. Wittmaack Eds., Springer, Berlin, 1991.

[16] T. Wagner, J.Y. Wang, S. Hofmann, Sputter Depth Profiling in AES and XPS, in: Surface Analysis by Auger and X-ray Photoelectron Spectroscopy, D. Briggs, J.T. Grant Eds., IM Publications, Chichester, 2003.

[17] S. Hofmann, Ultimate depth resolution and profile reconstruction in sputter profiling with AES and SIMS, Surf. Interface Anal., 30 (2000) 228–236.

[18] S. Hoffman, A. Zalar, Depth profiling with sample rotation: Capabilities and limitations, Surf. Interface Anal., 21 (1994) 304–309.

[19] R.M. Bradley, Theory of improved resolution in depth profiling with sample rotation, Appl. Phys. Lett., 68 (1996) 3722–3724.

[20] S. Hofmann, Sputter depth profiling of thin films, High Temp. Mater. Proc., 17 (1998) 1–27.

[21] NPL Sputtering Yields for Argon, www.npl.co.uk/mass-spectrometry/secondary-ion/sputter-yield-values, n.d.

[22] Angstrom Science, Elements and Compounds Sputtering Yields at 600 eV, www.angstromsciences.com/sputtering-yields, n.d.

[23] D.R. Baer, M.H. Engelhard, A.S. Lea, P. Nachimuthu, T.C. Droubay, J. Kim, B. Lee, C. Mathews, R.L. Opila, W.F. Saraf, W.F. Stickle, R.M. Wallace, B.S. Wright, Comparison of the sputter rates of oxide films relative to the sputter rate of SiO2, J. Vac. Sci. Technol. A, 28 (2010) 1060.

[24] H. Viefhaus, K. Hennesen, M. Lucas, E.M. Miller-Lorenz, H.J. Grabke, Ion sputter rates and yields for iron-, chromium- and aluminium oxide layers, Surf. Interface Anal., 21 (1994) 665.

[25] J. Bosgra, L.W. Veldhuizen, E. Zoethout, J. Verhoeven, R.A. Loch, A.E. Yakshin, F. Bijkerk, Interactions of C in layered Mo-Si structures, Thin Solid Films, 542 (2013) 210–213.

[26] O.A. Baschenko, V.I. Nevedov, Depth profiling of elements in surface layers of solids based on angular resolved X-ray photoelectron spectroscopy, J. Electron Spectrosc. Relat. Phenom., 53 (1990) 1–18.

[27] G.J. Tyler, D.G. Castner, B.D. Ratner, Regularization: A stable and accurate method for generating depth profiles from angle-dependent XPS data, Surf. Interface Anal., 14 (1989) 443–450.

[28] R. Jisl, Restoration of the depth-concentration profile from the angle-resolved relative intensities of X-ray photoelectron spectra, Surf. Interface Anal., 15 (1990) 719–726.

[29] M. Pijolat, G. Hollinger, New depth-profiling method by angular-dependent X-ray photoelectron spectroscopy, Surf. Sci., 105 (1981) 114–128.

[30] T.D. Bussing, P.H. Holloway, Deconvolution of concentration depth profiles from angle resolved X-ray photoelectron spectroscopy data, J. Vac. Sci. Technol. A, 3 (1985) 1973.

[31] O.A. Baschenko, M.A. Tyzykhov, V.I. Nefedov, Determination of depth-profiles in surface layers of solids by angular resolved X-ray photoelectron spectroscopy, Fresenius' J. Anal. Chem., 341 (1991) 597–600.

[32] V.I. Nefedov, O.A. Baschenko, Relative intensities in ESCA and quantitative depth profiling, J. Electron Spectrosc. Relat. Phenom., 47 (1988) 1–25.

[33] W.A.M. Aarnink, A. Weishaupt, A. van Siltbout, Angle-resolved X-ray photoelectron spectroscopy (ARXPS) and a modified Levenberg-Marquardt fit procedure: A new combination for modeling thin layers, Appl. Surf. Sci., 45 (1990) 37–48.

[34] J.F. Watts, J.E. Castle, S.J. Ludlam, The orientation of molecules at the locus of failure of polymer coatings on steel, J. Mater. Sci., 21 (1986) 2965–2971.

[35] L.B. Hazcll, A.A. Rizvi, I.S. Brown, S. Ainsworth, The use of X-ray photoelectron take-off-angle experiments in the study of Langmuir-Blodgett films, Spectrochim. Acta B, 40 (1985) 739–744.

[36] C.E. Shannon, A mathematical theory of communication, Bell Syst. J., 27 (1948) 623–656.

[37] J. Skilling, Gull S.F., The maximum entropy method in image processing, IEEE Proc. F, 131 (1984) 646–659.

[38] A.K. Livesey, G.C. Smith, The determination of depth profiles from angle dependent XPS using maximum entropy data analysis, J. Electron Spectrosc. Relat. Phenom., 67 (1994) 439–461.

[39] R.W. Paynter, Modification of the Beer-Lambert equation for application to concentration gradients, Surf. Interface Anal., 3 (1986) 186–187.

[40] G.C. Smith, A.K. Livesey, Maximum entropy: A new approach to non-destructive deconvolution of depth profiles from angle-dependent XPS, Surf. Interface Anal., 19 (1992) 175–180.

[41] J. Skilling, R.K. Bryan, Maximum image reconstruction: General algorithm, Mon. Not. R. Astron. Soc., 211 (1984) 111–124.

[42] M. Olla, G. Navarra, B. Elsener, A. Rossi, Nondestructive in-depth composition profile of oxy-hydroxide nanolayers on iron surfaces from ARXPS measurement, Surf. Interface Anal., 38 (2006) 964–974.

[43] Y. Yonamoto, Application of maximum entropy method to semiconductor engineering, Entropy, 15 (2013) 1663–1689.

[44] M. Szklarczyk, K. Macak, A.J. Roberts, K. Takahashi, S. Hutton, R. Glaszczka, C. Blomfield, Sub-nanometer resolution XPS depth profiling: Sensing of atoms, Appl. Surf. Sci., 411 (2017) 386–393.

[45] C.R. Smith, G.J. Erickson, Maximum-Entropy and Bayesian Spectral Analysis and Estimation Problems, Proceedings of the Third Workshop on Maximum Entropy and Bayesian Methods in Applied Statistics, Wyoming, 1–4 August 1983, Springer, Dordrecht, 1987.

[46] S.F. Gull, J. Skilling, Quantified Maximum Entropy MemSys5 Users' Manual, Copyright c MEDC, Edmunds, 1999.

6 Miscellanea

6.1. GRAPHICS

RxpsG uses a modified plot() function which enables plotting automatically the sequence of spectra of a given XPS Sample or the single core-line with the results of the analysis performed namely *Baseline*, *Fit Components*, and *Best Fit*. The plot() function also recognize if noise filtering, differentiation, and/or analysis of the *Valence Band* were performed to correctly display the results. This renders easy and automatic the visualization of the spectral data just typing plot (*XPS_Sample_name*) or plot(*Core-Line_name*). However, the simple plot() function cannot be directly used when customized plots must be generated. In this case the user should set the long list of parameters (just type *par()* to display the graphic parameters applied to plot the data). To render the selection of the various options more user friendly, four different functions were implemented to satisfy the operator needs. They are respectively the *Custom Plot*, *Overlay spectra*, *Compare spectra*, and *Two Y-Scale Plot* options which will be described in the next sections.

6.1.1. The *Custom Plot* Option

The *Custom Plot* function is designed to progressively build the plot step by step selecting the desired graphical options. This makes the composition of the figure easy and straightforward. The *Custom Plot* GUI is reported in Figure 6.1. The GUI is in form of notebook with six pages each dedicated to a specific part of the figure. The first page regards the control of the axes. At the top are the drop-down menus to select the XPS Sample among those loaded in *RxpsG*, and the core-line to plot. Upon selection, the correspondent spectrum is visualized. On the top-right of the GUI, the user can select if direct or reversed x-scale must be drawn. By default, the scale is reversed, since spectra are generally plotted versus binding energies.

The same page groups also the options regarding the axes style. The item *Ticks* enables plotting ticks on the desired couple of axes. The default is the *bottom* X and *left* Y axes. Other options are *top* and *right* axes, ticks on all the four axes or *custom* X, *custom* Y as shown in Figure 6.1b. Selection of one of these two latter cases opens a window where the user defines the minimum and maximum values of the axis scale, the number of desired ticks, and where to start ticking.

In the middle of the *Custom Plot* GUI, the user finds the *Scale* item. The drop down menu shows the possible scale styles that can be applied to the X and Y axes. The *Standard* format is the linear X, Y scales. Logarithmic and different power formats are also available (see Figure 6.1c).

Just next on the right, the user finds the option *Axis Label Orientation* to adapt scale numbers in the optimal way on the plot. Figure 6.2 shows the effect of an *Axis Label Orientation* at 45°. Figure 6.2 also shows that the default title "C 1s" was changed in "Virgin Sample C1s".

DOI: 10.1201/9781003296973-6

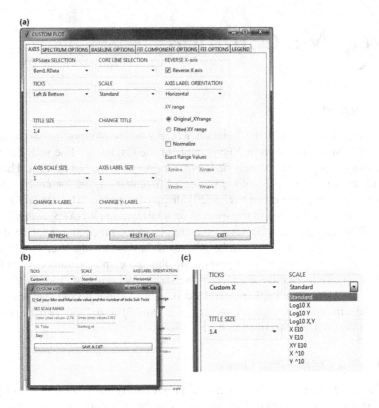

FIGURE 6.1   (a) The *Custom Plot* GUI. (b) Window enabling customization of X, Y axes ticks. (c) Different scale types are available for the X and Y axes.

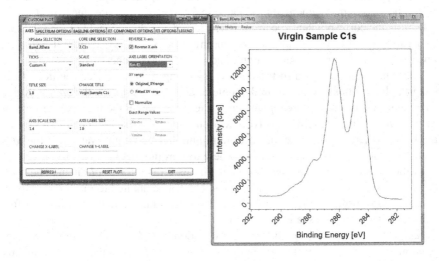

FIGURE 6.2   The *Custom Plot* GUI allows changing title and axis labels, their size, and orientation.

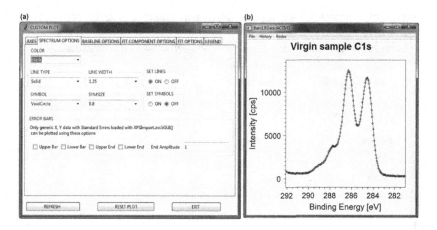

FIGURE 6.3   (a) Second page of the *Custom Plot* GUI. (b) Spectral rendering, both *solid line* and *symbols* are used to plot the spectrum.

In addition, the axis numbers and label dimensions were increased to render them sufficiently clear for a publication considering that generally the published figures are pretty small. In this first notebook page, the user can also change the *X, Y labels* if the default units are not satisfactory. Finally, the plotting region can be restricted to the *RegionToFit* to better visualize the spectral fit or normalize the spectrum or exact *X min/max, Y min/max* values can be entered.

The second page of the *Custom Plot* GUI regards the spectrum as visualized in Figure 6.3. Here, the user can select the color for plotting the spectrum; by default, it is black. In this page, it is also possible to define the plotting style, if *lines* or *symbols* or both are used to plot the data. The user can select different line patterns or symbols and their width and dimension.

Errors are commonly estimated applying a statistics to the experimental data. This is generally not performed when photoelectron spectra are acquired. The data provided by an XPS instrument are generated by applying a certain integration time and a number of sweeps. A final error can be estimated but generally it is assumed that it is small for good SNR. However, it is possible to use *RxpsG* to manipulate data from other sources provided they are converted in *Ascii* textual format. As illustrated in Section 2.3.3, it is possible to load *Ascii* data which can be coupled with their statistical error. In this case, it is possible to plot the error bars just marking the checkboxes to add to the upper/lower part of the error bars. The option *End Amplitude* is used to modify the length of the error bar ends. The *Baseline, Fit Components*, and *Fit* pages are very similar to the previous one. Just as an example, Figure 6.4 reports the *Fit Component* page. Again the user can select the color to plot the fitting components, the plotting style, if lines, symbols, or both must be used. An additional option regards the possibility to add (the default) or not the fit component name ($C_1 \ldots C_n$) to the plot.

Finally, the last page of the *CustomPlot* GUI is dedicated to the legend and is illustrated in Figure 6.5. At the top of the GUI, a check-box enables the legend which,

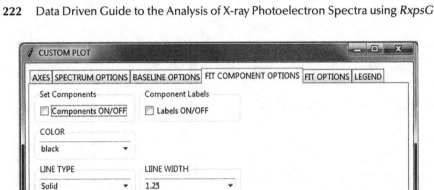

FIGURE 6.4   the *Fit Component* page of the *Custom plot* GUI.

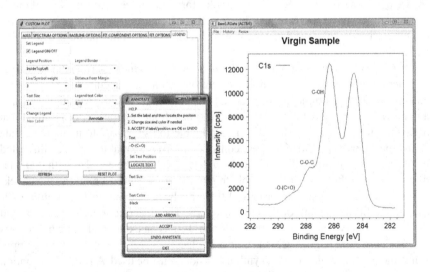

FIGURE 6.5   On the left: the *Legend* notebook page of the *Custom Plot* GUI. In the middle: the window appearing when the *Annotate* button is pressed. Finally on the right, the legend added in the *Inside top-left* position and the annotated C 1s spectrum showing the bond assignment.

by default, will appear below the Figure title. The drop-down menu allows the user to place the legend in pre-defined different positions. It is possible to modify these positions by selecting the distance from the figure margin. Other options encompass: legend border to plot a box around the legend; line/symbol weight to increase the thickness or dimension of line/symbol used for plotting the spectrum; size of the legend text. By default, the legend corresponds to the spectrum name, C 1s as in Figure 6.5. It is possible to modify the legend text simply entering the new text. Finally it is possible to add annotations to the plotted figure. When the annotation button is pressed a new window opens (see Figure 6.5) where the user can define the annotation text, the text size, the position. If a label is placed on the Figure in a wrong position, the UNDO button cancels the last operation. Figure 6.5 shows the legend and the annotation made adding the bond assignment for the C 1s core-line. When the available space is limited and annotation becomes difficult, it is possible to add an arrow using the *Add Arrow* button to relate annotation text to a specific spectral feature.

6.1.2. THE *OVERLAY SPECTRA* OPTION

As the name says, *Overlay Spectra* allows the user to overlap and compare spectra. In this, options are similar to that of the *Custom Plot*. Also in this case, the GUI is organized in a notebook resembling that of the *Custom Plot*. Figure 6.6a shows the first page of the *Overlay Spectra* listing all the XPS Sample data files loaded in *RxpsG*. The user starts selecting the desired XPS Sample. Immediately, the list of core-lines composing the selected data file will appear in the graphical window. The user must select the core-line to compare which will be added to the list displayed in the bottom part of the GUI by pressing the button '*Save Selection*' (see Figure 6.6a).

Let us suppose to compare the C 1s core-line. The user will select the C 1s core-line for each of the XPS Samples listed in the GUI and pressing the *Save Selection*

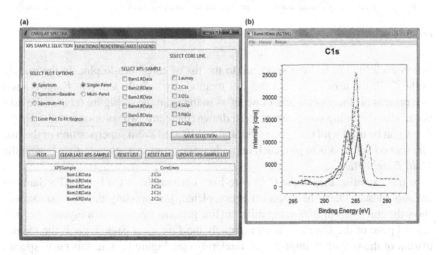

FIGURE 6.6 (a) The *XPS Sample Selection* page of the *Overlay Spectra* GUI. (b) C1s core-lines from different XPS Samples. The different spectra are drawn using different line patterns.

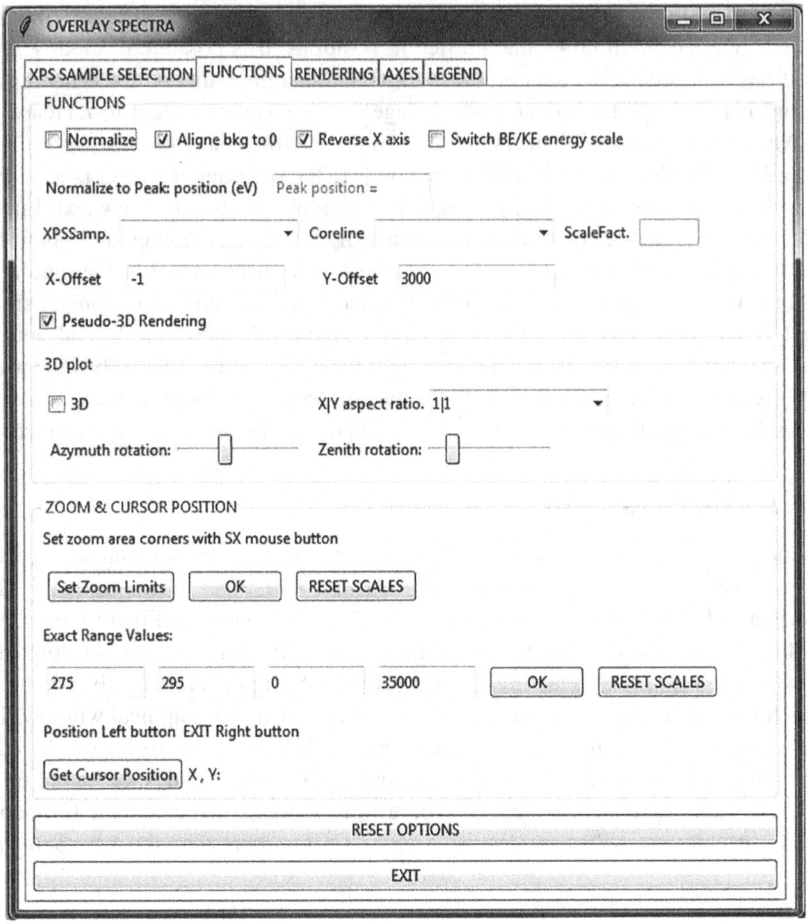

FIGURE 6.7   The *Functions* page of the *Overlay Spectra* GUI.

button each time to add the spectrum to the list of core-lines to plot. In the middle of the GUI, the user can select the plot method. By default, a *Single Panel* is used to represent all the spectra in a drawing as in the example of Figure 6.6b. If the user needs also the fitting components to be drawn, there are two solutions: still a *Single Panel* can be used with the spectra vertically shifted to avoid superposition of the fits. The second solution is to plot each individual spectrum in a separate panel using the *Multi Panel* option.

In this example, the simple C 1s core-lines deriving from the listed XPS Samples are superposed. Once the selection is completed, just pressing the *Plot* button will show the core-lines in B/W using different line patterns as reported in Figure 6.6b. The second page of the *Overlay Spectra,* namely the *Functions* page, regards the manipulation of the spectra to improve the rendering (see Figure 6.7). In this page, spectra can be aligned to zero, eliminating the background, or normalized, shifted in X and Y direction to create a waterfall effect for a pseudo-3D representation (see Figure 6.8a).

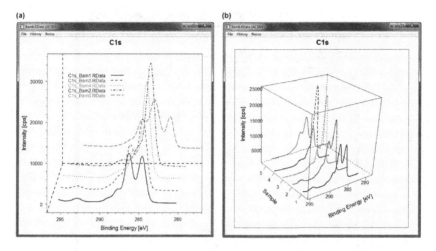

FIGURE 6.8 (a) Pseudo-3D generated by the option selection shown in Figure 6.7. (b) Real 3D representation of spectra list of selected spectra.

The X and Y range can be adapted to optimize the figure considering the applied shifts. The data can be plotted also in a real 3D fashion as shown in Figure 6.8b.

The page *Rendering*, illustrated in Figure 6.9, contains the options related to colors, line patterns, and symbols. Selecting the *Rainbow* mode, spectra are plotted in colors while the B/W uses black lines with different patterns. The user can select *Set Lines ON/OFF* or *Set Symbols ON/OFF* to plot spectra using lines or symbols or both. Line width and symbol size can be changed following the own needs. Color can be personalized but this does not affect the default list of colors. Combination of colors, different pattern and symbols can be utilized to optimize the rendering emphasizing the differences among spectra.

Finally on the left side of the page, the used list of colors for drawing the spectra and the fit elements is shown (Baseline, Fit components, and Best Fit). The user can freely modify the list of colors selecting from the color chart provided by R. These changes, however, will affect only the data in the actual plotting session and would not change the default settings. To change the default list of colors following the personal needs, the option *Preferences* must be used as described in Section 6.5.

The *Axes* page of the *Overlay Spectra* GUI enables the user to modify the axes style. The page is similar to that already described for the *Custom Plot* GUI and is shown in Figure 6.10.

On the top-left, it is possible to select the couple of axes where to draw the ticks. The default is *Left-Bottom*. *Top-Right*, *Both* or *Custom X*, *Custom Y* are also possible. In the case of custom axis selection, the user must enter the number of ticks to draw and the min and max values of the axis scale. See also the previous section. Follow the *X scale* and *Y scale* items to modify the relative axis style. Available options are *Standard* (the default), *Power*, *Log10*, *Ln* (natural logarithm), *value^10* for both the X and Y scales. Just below are the drop down menu to modify the title, the axes numbers, and their labels. If 3D plot is enabled, the user can also modify the label of

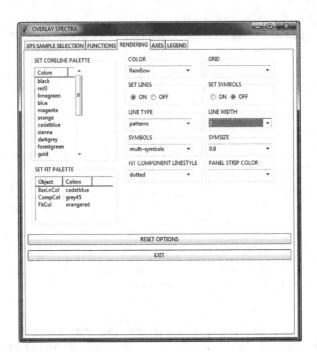

FIGURE 6.9   The *Rendering* page of the *Overlay Spectra* GUI.

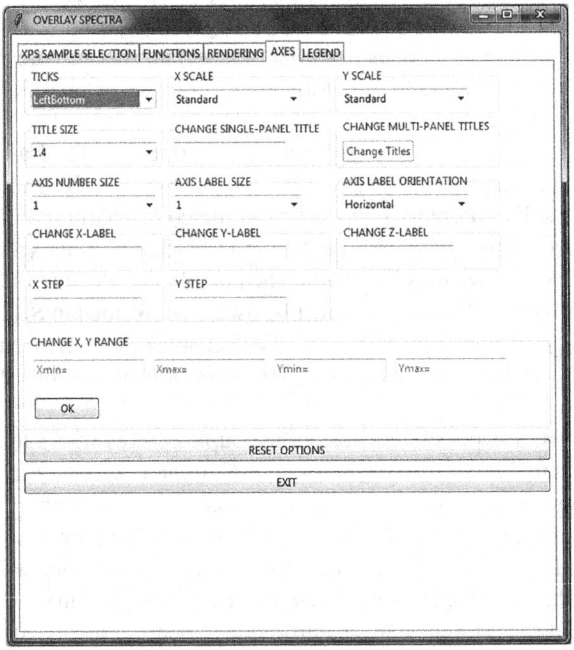

FIGURE 6.10   The *Axes* page of the *Overlay Spectra* GUI.

FIGURE 6.11 Layout of the *Legend* page of the *Overlay Spectra* GUI.

the Z axis. In the case of selection of the *Multi Plot* mode, the title of each panel can be modified. Finally to optimize the figure rendering, the user can change both the X and Y scale steps, and *RxpsG* will adapt the scale to the selected values. It is also possible to enter the axes min/max values to refine the plotted region. The last page of the notebook regards the plot *Legend* (see Figure 6.11).

At the top, the user enables the *Legend* and the extension of the default core-line name with the parent XPS Sample name. Follow the options concerning the *Legend* position, the possibility to organize the legends in a matrix with the desired number of columns, the *Legend* text size and distance of the *Legend* from the plot margin, the width and dimension of *Legend* symbols. The effect of the selections made in the *Legend* page of Figure 6.11 is shown in Figure 6.8a.

6.1.3. The *Compare XPS Samples* Option

The *Compare XPS-Samples* option compares core-lines from different XPS Samples as the *Overlay Spectra* does. The difference between the two procedures stands in the different way to compare spectra. In *Overlay spectra*, single core-lines selected from different XPS Samples are grouped and overlapped in single plots. In *Compare XPS-Samples*, the user selects the XPS Samples of interest and automatically the common core-lines are visualized. The user can select which of the core-lines to plot. Now a panel is dedicated to each group of core-lines corresponding to the same chemical element but coming from different XPS Samples. The user can then compare how spectra change from sample to sample.

Also the *Compare XPS-Samples* option is structured in pages. The first one is represented in Figure 6.12.

As illustrated in Figure 6.12a, the list of XPS Samples loaded in the *RxpsG* is visualized. The user can freely select which of these data-samples desires to visualize and which of the core-lines present in all the selected XPS Samples must be plotted. In the present example, all the loaded XPS Samples are chosen as well as all the common core-lines, namely *Survey*, *C 1s*, and *O 1s*. Just pressing the *Plot* button the groups of core-lines of same type are overlapped in separated panels as reported in Figure 6.12b and a legend explaining the correspondent samples is added. The second page of the *Compare XPS-Samples* is reported in Figure 6.13.

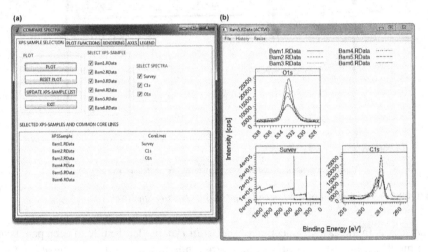

FIGURE 6.12   (a) The main page of the *Compare XPS-Samples* GUI. A number of XPS Samples are selected: the common core-lines, Survey, C1s, and O1s. (b) These core-lines are visualized in separated panels in B/W style.

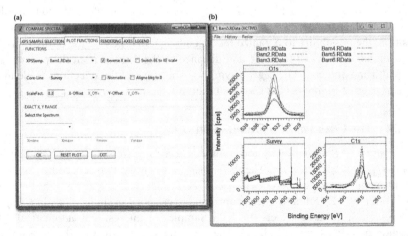

FIGURE 6.13   (a) The second page of the *Compare XPS-Samples* GUI. In (b), the effect of the application of the multiplication for the rescaling factor 0.3 applied to the *Survey* spectrum of the *Bam1.RData* (compare with the same panel in Figure 6.12b).

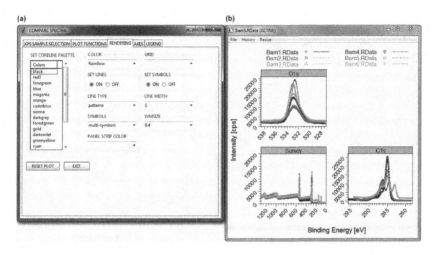

FIGURE 6.14 (a) *Plot Functions* page. In (b), the effects of the selections done in the page of *Compare Spectra* shown in (a): both patterned lines and different symbols are used to plot spectra.

The possible operations which can be applied to the groups of core-lines are the same of those already described in the *Custom Plot* and *Overlay Spectra* options. Essentially the X-axis, which by default is Binding Energy, can be reversed. The Binding Energy scale can be transformed in Kinetic Energy for all the groups of core-lines. On the left of the page displayed in Figure 6.13(a), the user can select the XPS-Sample and the core-line on which apply a specific operation. In Figure 6.13(a), the *Bam1.RData* and the *Survey* spectrum are selected and a rescaling factor equal to 0.3 is applied. *Normalization* and *Align Background to 0* as well as an exact X and Y ranges can be applied to a selected group of core-lines. The *Rendering* page in Figure 6.14 resembles that of the *Custom Plot* and *Overlay Spectra* GUIs.

Essentially the user can select if B/W or colors, lines, or symbols must be used to plot the data. Line width and symbol size can be modified and the palette used for plotting data can be personalized. In the *Axis* page are organized options related to the axis settings and is illustrated in Figure 6.15. Again the page size is similar to those described for the *Custom Plot* and *Overlay Spectra* functions. In the top area of the page, the user can select the scale style for the X and Y axes. "*Standard*", power scale, logarithmic (base 10 and natural), and scientific exponent formats are available. Custom X and Y increments can be defined to optimize the number of axis numbers with respect to the relative axis range. Also the orientation of axis numbers is available for both the axis.

The user can change the title and the axis labels and select their size. In the example of Figure 6.15, scientific exponent format for the Y axis, 45° orientation and custom X and Y steps, and increased X and Y axis label size were selected to produce the effects reported in Figure 6.15b.

The last page of the *Compare XPS-Samples* GUI regards the legends. By default, a legend is added to the plot to facilitate the identification of the origin of the spectral data. As shown in Figure 6.16, the *Labels* page contains a list of options to optimize the description of the spectral data.

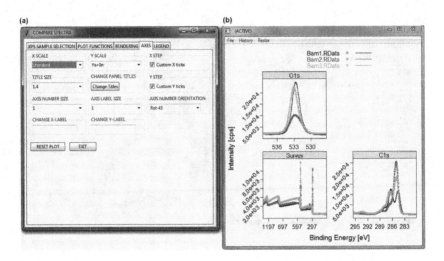

FIGURE 6.15 (a) The *Axis* page of the *Compare XPS-Samples* GUI. (b) The effect of the selected options shown in (a).

FIGURE 6.16 The *Legend* page of the *Compare XPS-Samples* function.

The user can enable/disable the presence of legends to add the *XPS-Sample* names to associate the spectra to the experiment. The legends can be ordered in one or more columns while the size and the text of the legends can be freely changed. Finally the user can increase the width of the legend lines or symbol size.

6.1.4. THE *TWO SCALE PLOTS* OPTION

This option is intended to help the user for plotting data in a two scale fashion. This option can be useful to plot data created by spectral analysis as a function of the experimental conditions. The user can plot the trend of an element concentration as a function of an experimental parameter and at the same time the variation of a physical entity describing the surface property of the analyzed samples. For example, for a given set of samples annealed at different temperatures, the user can plot the trend of the oxygen concentration as a function of the temperature and at the same time the wettability of their surfaces. At this aim, the user can organize the data in columns of a matrix and save them in a simple textual (*Ascii*) file. Then data can be loaded in *RxpsG* using the *Import Ascii* option (Section 2.3.3). As for the previous functions, also the *Two Y-Scale Plot* layout is in form of a notebook. The first page is shown in Figure 6.17.

The user must select the source file names, the *XPS-Sample 1* and the *XPS-Sample 2*, which can correspond to the same name as shown in Figure 6.17 if the data to plot are contained in just one file. Then one has to choose which are the *Data-set 1*

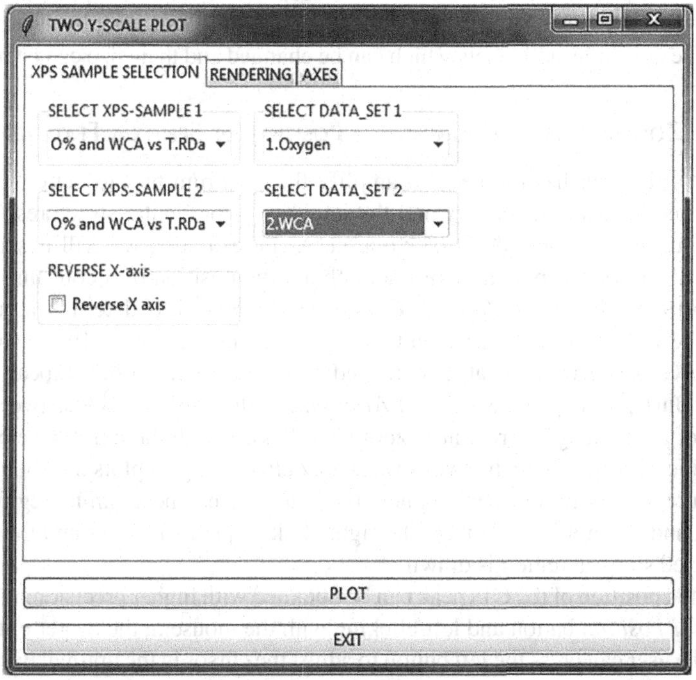

FIGURE 6.17   The first page of the *Two Y-Scale Plot* GUI.

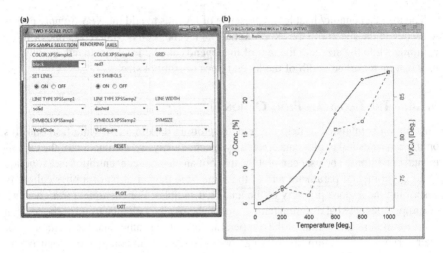

FIGURE 6.18   (a) The second page of the *Two Y-Scale Plot* option is dedicated to the rendering. (b) The effect of the selected options.

and the *Data-set 2* to plot and if the X-axis has or not to be reversed. The next pages contain options to personalize the plot. Figure 6.18a shows the second page where the user can select the B/W or colors, the line patterns, if symbols are or not utilized. The effect of the selected options is reported in Figure 6.18b. As it can be seen, both lines with different patterns and symbols can be used. If the rainbow option is used, the Y-axis assumes the same color of the correspondent set of data. Finally, the third page regards the axis, title, and labels which can be changed and their size can be modified.

6.1.5. Zooming and Obtaining the Position of Spectral Features

Frequently the user has the need to amplify the spectrum plot to better appreciate features and measure the energy and the intensity of particular structures. This can be made using the *Zoom & Cursor* option. As an example, we will describe how to make a zoom of a spectral region and obtain the position of a core-line peak. In Figure 6.19a is shown the *Zoom & Cursor* panel; the task is determining the position of the C 1s peak from an HOPG sample shown in Figure 6.19b. As it can be seen, the C 1s spectrum is rather extended and the estimation of the peak position difficult. Just pressing the *Set Zoom Area* button the user can select two opposite corners of the rectangular region to zoom by clicking with the left mouse button on the graphic window. Once the two corners are defined, *RxpsG* plots the zooming area (see Figure 6.19b). By left clicking near the blue crosses, the zooming region can be adapted, and when it is optimized, the right click stops the process and the zoom of the selected spectral region is drawn.

Then the position of the C1s peak can be obtained with higher precision by pressing the *Cursor Position* button and left-clicking with the mouse in the center of the peak. Again, the user can press the left button to adjust the cursor in the optimal position and to get the binding (kinetic) energy of the spectral features of interest. The marker position and the correspondent X, Y coordinates are visible at the bottom of the *Zoom &*

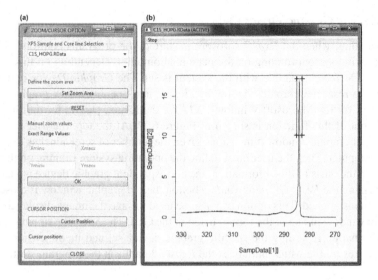

FIGURE 6.19 (a) Layout of the *Zoom & Cursor* option. (b) C 1s from an HOPG sample. Indicated is the region to zoom.

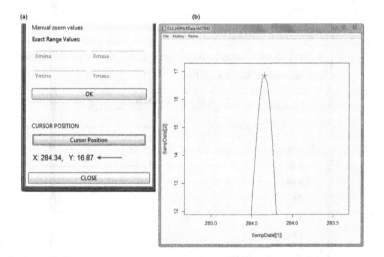

FIGURE 6.20 (a) The *Cursor Position* button activates reading the cursor position. (b) By left clicking on the spectrum, a cross is plotted in the cursor position and the correspondent X, Y coordinates reported in the GUI (see arrow in panel A). The X, Y coordinates are updated when successive clicks are made.

Cursor GUI as indicated by the arrow in Figure 6.20a. Their value is updated when successive clicks are made. Reading coordinates is stopped by right-clicking.

Finally, it is also possible to manually enter the exact values of *Xmin*, *Xmax*, *Ymin*, and *Ymax* relative to the region to zoom when the precise ranges of the X and Y scales are required.

6.1.6. THE *GRAPHIC DEVICE* OPTION

At beginning, when the *RxpsG* software is started, the programs controls which is the operating system running on the personal computer. Depending on this latter, a *Graphic Device* compatible with the system is set. The *Graphic Device* option can be used to reset the graphic window by closing and opening a new graphic device (*Windows* for the Microsoft Windows, *X11* for the Linux, *Quartz* for the Mac-OS). The layout of this function is shown in Figure 6.21. At the top, the user can select the default graphic window dimensions given the number of screen inches of his personal computer. Then the user must define the operating system running on the computer. On the basis of this information, the correspondent graphic device is identified and pressing the *Reset Graphic Window* button the old graphic window is closed and a new one is opened. This option can also be used to save the image displayed in the graphic window in one of the format listed in the GUI. The *Save & Exit* button stores the information regarding the specific graphic device set and its dimensions in the *XPSSettings.ini* file which initializes the *RxpsG* defaults at the program start.

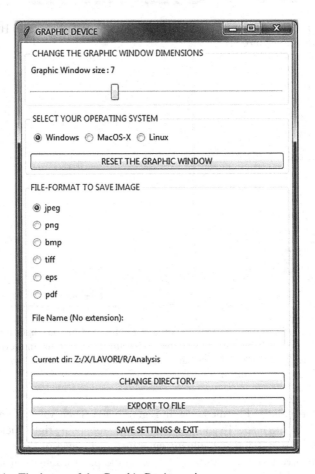

FIGURE 6.21   The layout of the *Graphic Device* option.

6.2. RECOVERING EXPERIMENT INFORMATION

Interpretation of the results of the spectral analysis requires the knowledge of the experimental conditions and in particular which are the synthesis conditions of the analyzed samples and the settings used for the acquisition of the spectra. This information can be obtained using the *XPS Sample Info* option (under the *Info & Help* of the main menu) which provides experimental details concerning the analyzed sample and the acquisition conditions for each of the core-lines contained in the *XPS Sample*. As an example in Figure 6.22, it is shown the layout of the *XPS Sample Info* GUI.

At the top of the GUI layout, information about the experimental session is reported. In particular are given information about the research project for which the sample is analyzed, and about the samples' preparation/synthesis (these notes have to be saved during the acquisition of the spectra), the operator making the acquisition, and the names of the acquired core-lines. This information is editable and the changes will be saved in the XPS Sample when the *Save & Exit* button is pressed.

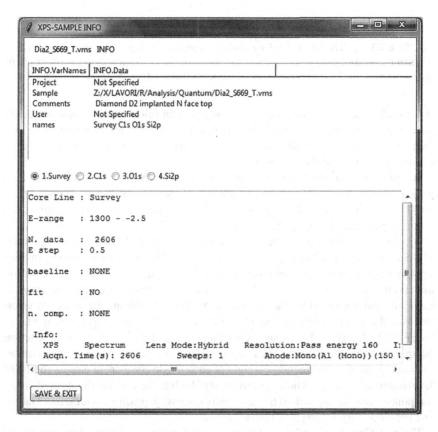

FIGURE 6.22   The *XPS Sample Info* GUI. On the top is listed information regarding the data file and the analyzed sample. Each of the core-lines of the *XPS Sample* is listed in the middle of the GUI. By clicking on one of these core-lines, the correspondent acquisition conditions are visualized.

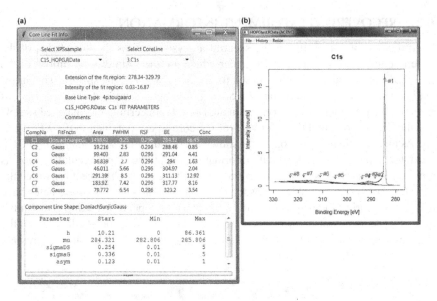

FIGURE 6.23 (a) The *XPS Core-Line Fit Info* layout. On the top of the panel is reported the information regarding the selected core-line shown in (b). In the middle is the list of the fitting component and on the bottom is the fit parameters describing the selected fit component.

In the middle are listed the core-lines. The user can select each of them to see the correspondent acquisition conditions which are reported at the bottom of the GUI. Changes made in this textual window are not saved in the XPS Sample at exit, since it is not possible to modify the acquisition conditions.

The previous option shows the properties of the XPS Sample and its core-lines as they were acquired. Information regarding the spectral analysis performed on the core-lines can be retrieved using the *Core-Line Fit Info* option. The layout of this option is illustrated in Figure 6.23 for a typical core-line fit.

When the XPS Sample and the core-line are selected (by default they are the actual XPS Sample and the actual core-line), the *Core-LineFit Info* provides the information regarding the analysis performed on the actual core-line. Details about the core-line are reported at the top of the GUI: the *Region To Fit* extension and core-line intensity range, the *Baseline* used for background subtraction. In the middle is a table where are listed the set of fitting component used. For each of these components, the *Core-Line Fit Info* describes the essential characteristics as the correspondent fit-function, the spectral area, the FWHM, the assigned sensitivity factor, and position. The last column represents the percentage weight of each single component considered 100% the spectral area of the whole spectrum. By double clicking on the desired fitting component, the correspondent fit parameters (intensity, position, width, mix Gaussian-Lorentzian, asymmetry, etc.) are obtained.

The *XPS Core-Line Fit Info* can be used also for non-conventional processing. As an example, as it appears in Figure 6.24 a, the *XPS Core-Line Info* gives information regarding the fitting components used to find the VB_{top} position (see Section 3.9.5). In this case, however, the fit acts as a sort of filter allowing a good estimate of VB_{top}

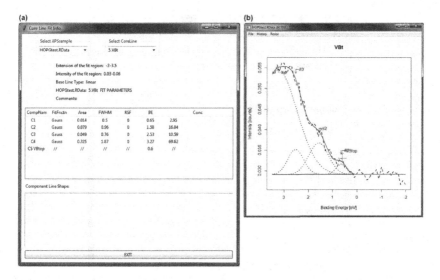

FIGURE 6.24 (a) Layout of the *XPS Core-Line Fit Info* in the case of the estimation of the VB_{top} made via spectral fitting. (b) The correspondent VB spectral fit.

as shown in Figure 6.24b. The only requirement is the quality of the fit which must be very good. The number, position, and FWHM of the components are not relevant. For this reason, clicking on the spectral component of the table, no further information is provided.

6.3. THE *INTERPOLATION* AND *DECIMATION* OPTIONS

Sometimes it happens that specific processing methods require acquisition of the core-lines using a smaller energy step so that their precision increases. However, there are cases in which the energy step and the energy scale of two spectra must be exactly the same. For example, this occurs when we desire to deconvolve the loss feature from a spectrum as described in Section 3.8.8 for the analysis of the Carbon Auger spectrum. The option *XPS Interpolation Decimation* can be used to change the energy step by adding or removing points from the spectrum. Figure 6.25 shows the *XPS Interpolation Decimation* layout. As usual, on the top there are two drop down menus to select the XPS Sample and the core-line to modify. At selection of this latter, the original energy step is visualized in bold (see Figure 6.25).

Pressing the button *Compute RxpsG* verify if the energy scale (and spectrum) of the selected core-line can be modified just applying a data *Interpolation* or a *Decimation* or if the new energy step must be obtained via successive application of these two operations.

The user can also directly apply an *Interpolation* or a *Decimation* just specifying the number of points to interpolate/decimate and then pressing the *Compute* button. Any time *RxpsG* gives information about the applied operation, namely the final number of data and energy step generated. The user can increase or reduce the

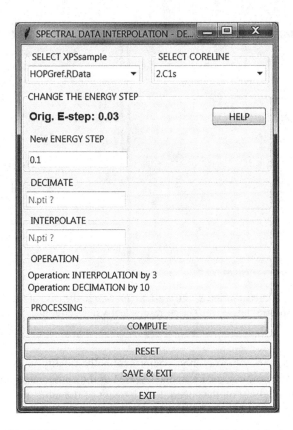

FIGURE 6.25   The *XPS Interpolation Decimation* GUI. In bold the original energy step of the C 1s core-line. Just below, in the *New Energy Step* window, is indicated the new energy step to set.

number of data of a spectrum at will, thus making the energy step of the core-line more dense or sparse upon his needs.

6.4.   *DATA SPRUCING* OPTION

It may occur that raw data are affected by acquisition errors as undesired spikes or null data. It is possible to correct the raw spectral data using the *Sprucing Up* option (under the *Analysis* main menu) whose layout is shown in Figure 6.26.

As shown in Figure 6.26b, the wide spectrum is obscured by the presence of a spike. Indicated is also the region containing the spike. The *Select Region* button must be pressed to define the two opposite corners of the region around the spike by left clicking the mouse on the spectrum. Reading the position is stopped by pressing the right button. An expanded plot of the selected region is presented, and the user can refine the extension of the region around the data to correct by clicking near the crosses at the corners. Again, the right mouse click stops reading the cursor X, Y coordinates. The button *Edit Region* is pressed to edit the correspondent

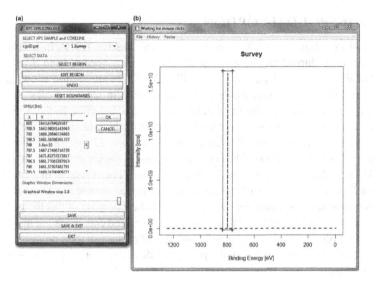

FIGURE 6.26 (a) Layout of the *Sprucing UP* GUI. (b) Typical example of spike affecting the spectrum.

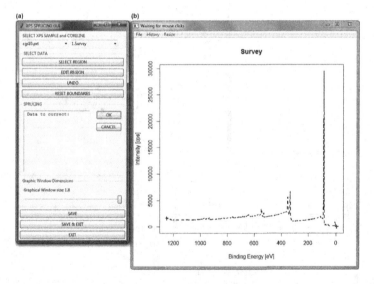

FIGURE 6.27 (a) After pressing the button *OK*, the edited region is reset, and (b) the corrected data are plotted. The software is ready to select another region to edit/correct.

spectral data. The user can manually scroll and correct the wrong values, and press *OK* to finalize the operation. The corrected spectrum will be plotted, Figure 6.27a, b, and data will be saved in the main *RxpsG* memory by pressing the buttons *Save* or *Save & Exit*.

6.5.   THE *SPLIT DATA* OPTION

The split data option is written for users of Scienta-Omicron ESCA Instruments. The acquisition procedure of that instrument allows acquiring spectra from diverse specimens but all the data are saved in just one file. The *Split Data* option allows selecting single core-lines of the original data file and save them in separated files. The layout of relative to this option is illustrated in Figure 6.28a, b. Selection of the XPS Sample will generate a check-box table showing the content of the original data file. Up to 64 core-lines can be listed in that table although only the first 12 spectra are plotted (Figure 6.28b). The user can choose the core-lines to save in the separated file; in the example of Figure 6.28a, the first five core-lines are selected. Then pressing the button *Select and Save Spectra*, a window will appear where the user can browse the destination folder and enter the name of the new data file. By default, the extension of this file is. RData.

Now pressing the *Clear Selections* button, the exported core-lines are marked with "=>" to distinguish them from the other spectra. The user can then proceed selecting other core-lines to save and repeat the procedure described earlier to store the spectra in a new file.

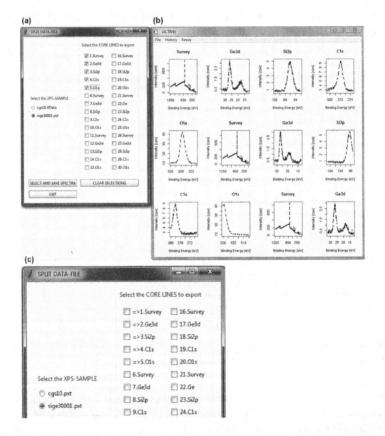

FIGURE 6.28   (a) Example of. PXT data file containing acquisitions on multiple samples. The selected core-lines will be saved in a separated file. (b) The first 12 core-lines from different samples. (c) The core-lines already saved in a file are marked with "=>".

6.6. *PREFERENCES* OPTION

The *Preferences* option enables the user to personalize some of the defaults of the *RxpsG* software. Essentially the user can set the graphic parameters. First of all, it is possible to define the operating system running on the personal computer to allow *RxpsG* to set the correct graphical device (graphic window) as shown in Figure 6.29 at the top. Then it is possible to define a personal palette for plotting the data. On the right part of the *Preference*, GUI is a table where the list of available colors, the line patterns, and symbols are listed in the order they will be used for plotting data (first core-line in black solid line and void-circle symbol, second core-line will be plotted using the red color, dashed line, void-square symbol, etc.) (see Figure 6.29). For colors, the user can refer to the color tables used by R with names, or the corresponding hexadecimal code are given.

Let us consider the various options of the *Preferences* GUI.

*   *Graphic Window size*: This sets the default dimension of the windows opened by other options as *XPSAnalysis, XPSMoveComponent, XPSExtract, XPSSprucing,* and *XPSVBTop*. Depending on the dimensions of the screen of your PC, you can increase/decrease the window dimension. The *Graphic Window size* can be set also using the *Set Graphic Device* option.
*   *Select the operating system for graphics*: The selection of the operating system defines the graphic device used to open the graphic windows used by *RxpsG*. Also the operating system can be set using the *Set Graphic Device* option.

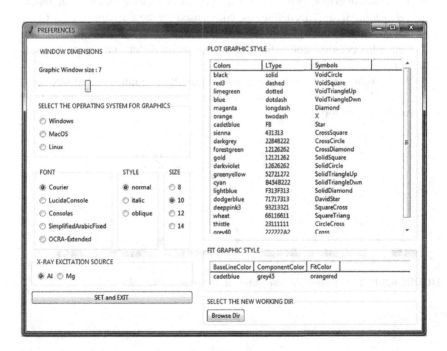

FIGURE 6.29   The *Preferences* GUI layout.

- *Font, Style, and Size*: defines the font, the style and the character dimension used in *RxpsG* outputs as those produced by *Quantify* and *Analysis Report*. Available fonts are the monospaced.
- *X-ray Excitation Source*: This defines the X-radiation energy used by the XPS-instrument. By default is the Al Kα, available is also Mg Kα. Knowledge of the excitation radiation is needed to switch from binding to kinetic energy in the graphical outputs.
- *Plot graphic style*: Here you can define the sequence of colors, line_styles, and symbols utilized by the plot options *Overlay Spectra, Compare Spectra, Custom Plot,* and *Two Y-scale Plot*. For a complete list of the available colors, visit the website: https://rstudio-pubs-static.s3.amazonaws.com/3486_7 9191ad32cf74955b4502b8530aad627.html
- You can change freely the sequence and the kind of colors used by the *RxpsG* to plot sequences of spectra to define your personal palette. Besides colors are indicated the available line patterns. The user can change the order of the line patterns or define his personal pattern using a string of digits of an hexadecimal number. Each digit corresponds alternatively to black and white segments which indicate the segment length. For example, the number *22848222* will correspond to the pattern shown in figure 6.30. Here, grey parts are white gaps.
- As for symbols, there are only 25 different shapes available in R. For a complete list of the available symbols, visit the website: https://r-charts.com/base-r/pch-symbols/
- It is possible to modify the order of the symbols following the own preference.
- *Fit graphic style:* This defines the colors used for the *Baseline*, the *Fit Components*, and the *best Fit* in the plot options. Also in this case, the user can define his personal colors for plotting the spectral elements mentioned earlier.
- *Set Working Dir:* This option allows setting the default working directory. Once *RxpsG* is opened, search for files to load starting from the default working directory.

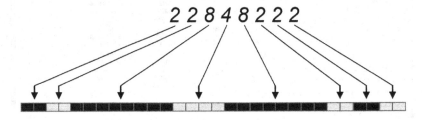

FIGURE 6.30   Line pattern corresponding to the code 22848222.

Index